BIOTOL

BIOTECHNOLOGY BY OPEN LEARNING

Analysis of Amino Acids, Proteins and Nucleic Acids

PUBLISHED ON BEHALF OF:

Open universiteit   and   **Thames Polytechnic**

Valkenburgerweg 167   Avery Hill Road
6401 DL Heerlen   Eltham, London SE9 2HB
Nederland   United Kingdom

BUTTERWORTH
HEINEMANN

Butterworth-Heinemann Ltd
Linacre House, Jordan Hill, Oxford OX2 8DP

 PART OF REED INTERNATIONAL BOOKS

OXFORD LONDON BOSTON
MUNICH NEW DELHI SINGAPORE SYDNEY
TOKYO TORONTO WELLINGTON

First published 1992

British Library Cataloguing in Publication Data
A catalogue record for this book is
available from the British Library

Library of Congress Cataloguing in Publication Data
A catalogue record for this book is
available from the Library of Congress

ISBN 0 7506 1502 8

Composition by Thames Polytechnic
Printed and Bound in Great Britain by
Thomson Litho, East Kilbride, Scotland

BOOKS IN THE BIOTOL SERIES

The Molecular Fabric of Cells
Infrastructure and Activities of Cells

Techniques used in Bioproduct Analysis
Analysis of Amino Acids, Proteins and Nucleic Acids
Analysis of Carbohydrates and Lipids

Principles of Cell Energetics
Energy Sources for Cells
Biosynthesis and the Integration of Cell Metabolism

Genome Management in Prokaryotes
Genome Management in Eukaryotes

Crop Physiology
Crop Productivity

Functional Physiology
Cellular Interactions and Immunobiology
Defence Mechanisms

Bioprocess Technology: Modelling and Transport Phenomena
Operational Modes of Bioreactors

In vitro Cultivation of Micro-organisms
In vitro Cultivation of Plant Cells
In vitro Cultivation of Animal Cells

Bioreactor Design and Product Yield
Product Recovery in Bioprocess Technology

Techniques for Engineering Genes
Strategies for Engineering Organisms

Principles of Enzymology for Technological Applications
Technological Applications of Biocatalysts
Technological Applications of Immunochemicals

Biotechnological Innovations in Health Care

Biotechnological Innovations in Crop Improvement
Biotechnological Innovations in Animal Productivity

Biotechnological Innovations in Energy and Environmental Management

Biotechnological Innovations in Chemical Synthesis

Biotechnological Innovations in Food Processing

Biotechnology Source Book: Safety, Good Practice and Regulatory Affairs

The Biotol Project

The BIOTOL team

OPEN UNIVERSITEIT, THE NETHERLANDS
Prof M. C. E. van Dam-Mieras
Prof W. H. de Jeu
Prof J. de Vries

THAMES POLYTECHNIC, UK
Prof B. R. Currell
Dr J. W. James
Dr C. K. Leach
Mr R. A. Patmore

This series of books has been developed through a collaboration between the Open universiteit of the Netherlands and Thames Polytechnic to provide a whole library of advanced level flexible learning materials including books, computer and video programmes. The series will be of particular value to those working in the chemical, pharmaceutical, health care, food and drinks, agriculture, and environmental, manufacturing and service industries. These industries will be increasingly faced with training problems as the use of biologically based techniques replaces or enhances chemical ones or indeed allows the development of products previously impossible.

The BIOTOL books may be studied privately, but specifically they provide a cost-effective major resource for in-house company training and are the basis for a wider range of courses (open, distance or traditional) from universities which, with practical and tutorial support, lead to recognised qualifications. There is a developing network of institutions throughout Europe to offer tutorial and practical support and courses based on BIOTOL both for those newly entering the field of biotechnology and for graduates looking for more advanced training. BIOTOL is for any one wishing to know about and use the principles and techniques of modern biotechnology whether they are technicians needing further education, new graduates wishing to extend their knowledge, mature staff faced with changing work or a new career, managers unfamiliar with the new technology or those returning to work after a career break.

Our learning texts, written in an informal and friendly style, embody the best characteristics of both open and distance learning to provide a flexible resource for individuals, training organisations, polytechnics and universities, and professional bodies. The content of each book has been carefully worked out between teachers and industry to lead students through a programme of work so that they may achieve clearly stated learning objectives. There are activities and exercises throughout the books, and self assessment questions that allow students to check their own progress and receive any necessary remedial help.

The books, within the series, are modular allowing students to select their own entry point depending on their knowledge and previous experience. These texts therefore remove the necessity for students to attend institution based lectures at specific times and places, bringing a new freedom to study their chosen subject at the time they need and a pace and place to suit them. This same freedom is highly beneficial to industry since staff can receive training without spending significant periods away from the workplace attending lectures and courses, and without altering work patterns.

Contributors

AUTHORS

Dr S.A. Boffey, Hatfield Polytechnic, Hatfield, UK

Professor A.M. James, 50 Brodrick Road, Tooting, London, UK

Dr R.J. Slater, Hatfield Polytechnic, Hatfield, UK

Dr J.M. Walker, Hatfield Polytechnic, Hatfield, UK

EDITOR

Professor A.M. James, 50 Brodrick Road, Tooting, London, UK

SCIENTIFIC AND COURSE ADVISORS

Dr M. C. E. van Dam-Mieras, Open universiteit, Heerlen, The Netherlands

Dr C. K. Leach, Leicester Polytechnic, Leicester, UK

ACKNOWLEDGEMENTS

Grateful thanks are extended to all who have contributed to the development and production of this book. In addition to the authors, editor and course advisors, special thanks go to Dr J. Gartland, Ms H. Leather, Ms J. Skelton and Professor R. Spier.

The development of this BIOTOL text has been funded by **COMETT, The European Community Action Programme for Education and Training for Technology**. Additional support was received from the Open universiteit of The Netherlands and by Thames Polytechnic.

Project Manager: Dr J. W. James

Contents

How to use an open learning text

An open learning text presents to you a very carefully thought out programme of study to achieve stated learning objectives, just as a lecturer does. Rather than just listening to a lecture once, and trying to make notes at the same time, you can with a BIOTOL text study it at your own pace, go back over bits you are unsure about and study wherever you choose. Of great importance are the self assessment questions (SAQs) which challenge your understanding and progress and the responses which provide some help if you have had difficulty. These SAQs are carefully thought out to check that you are indeed achieving the set objectives and therefore are a very important part of your study. Every so often in the text you will find the symbol Π, our open door to learning, which indicates an activity for you to do. You will probably find that this participation is a great help to learning so it is important not to skip it.

Whilst you can, as an open learner, study where and when you want, do try to find a place where you can work without disturbance. Most students aim to study a certain number of hours each day or each weekend. If you decide to study for several hours at once, take short breaks of five to ten minutes regularly as it helps to maintain a higher level of overall concentration.

Before you begin a detailed reading of the text, familiarise yourself with the general layout of the material. Have a look at the contents of the various chapters and flip through the pages to get a general impression of the way the subject is dealt with. Forget the old taboo of not writing in books. There is room for your comments, notes and answers; use it and make the book your own personal study record for future revision and reference.

At intervals you will find a summary and list of objectives. The summary will emphasise the important points covered by the material that you have read and the objectives will give you a check list of the things you should then be able to achieve. There are notes in the left hand margin, to help orientate you and emphasise new and important messages.

BIOTOL will be used by universities, polytechnics and colleges as well as industrial training organisations and professional bodies. The texts will form a basis for flexible courses of all types leading to certificates, diplomas and degrees often through credit accumulation and transfer arrangements. In future there will be additional resources available including videos and computer based training programmes.

Preface

Five decades have elapsed since Avery and his co-workers obtained the first experimental evidence that DNA carries the genetic message. The intervening years have witnessed a great increase in our knowledge of how this genetic information is translated into proteins and how proteins fulfil their various roles as mediators of metabolism (enzymes) and as agents of transport, defence and regulation. Along with these developments has grown increasing recognition that biological systems, or their products, are potentially of tremendous value in a large number of business sectors. Biological products and processes are of increasing importance in health care, agriculture, food processing, chemical synthesis and environmental management.

The emergence of modern biotechnology in the 1970s can be traced largely to the discovery of the enzymes which enable us to specifically splice and resynthesise DNA and to the development of techniques to analyse and process nucleic acid and proteins on both the micro- and macro- scales. Underpinning these developments in the biological sciences and biotechnology is a requirement to be able to extract, purify and analyse nucleic acids and proteins. This text aims to provide the knowledge that will enable the reader to develop experimental stratagies for, and to apply appropriate techniques to, the study of these important bio-molecules.

The text has been prepared by an author:editor team who have long-term experience in the delivery of courses on the techniques of protein and nucleic acid analysis for which they have received international recognition. They have used this experience to produce a readily accessible and logically developed learning resource. Although the main thrust of the text is practical in nature, they have underpinned the description of techniques with the essential knowledge that will enable the reader to identify the advantages, limitations and pitfalls that may be encountered with each technique. Care has been taken to ensure that data derived from the techniques are interpreted correctly and with the appropriate level of caution. The text, therefore, is not merely a collection of procedures but provides an opportunity to learn how to develop sound strategies for the isolation and analysis of these important bio-molecules.

The reader should realise that the material covered by this text has application in a wide range of biological sub-disciplines including investigations into the processes of intermediary metabolism and their regulation, studies on the physiological and pharmacological activities of proteins and other molecules, and exploration and manipulation of genetic information. The text, therefore deals with the techniques which are indispensable in many areas of research and exploitation of biological systems.

We point out that this text is partnered by two other BIOTOL texts. 'Techniques used in Bioproduct Analysis' examines general techniques used in the study of biomolecules. It includes descriptions of harvesting, disruption and sub-cellular fractionation of biological material as well as descriptions of the principles of the general techniques of separation, purification, structural analysis, electrometry and colorimetry. The present text extends this discussion by describing the specific application of these techniques to amino acids, proteins and nucleic acids. A third BIOTOL text, 'Analysis of Carbohydrates and Lipids', describes the application of analytical techniques to the structurally and metabolically important carbohydrates and lipids. Together, the three texts provide a very broad range of knowledge and skills required in the study of biological systems.

Our thanks go to all who have contributed to the preparation and production of this important study programme.

Scientific and Course Advisors: Professor M.C.E van Dam-Mieras
Dr C. K. Leach

Introduction - fundamental techniques

Introduction - fundamental techniques

In this book we shall consider the extraction, purification, estimation and general properties of amino acids, peptides, proteins and nucleic acids. It will be assumed that you have a basic understanding of the general principles underlying the disruption of biological material, separation methods and the methods available for analysis and structure determination. This knowledge may have been obtained from a study of the BIOTOL text, 'Techniques used in Bioproduct Analysis', from lectures or from practical experience in the laboratory. The first chapter of this book seeks to refresh your memory of the important points of these different topics and if you are unsure of any of these then you are advised to revise the appropriate section. We recommend the BIOTOL text, 'Techniques used in Bioproduct Analysis'.

1.1 Cell disruption

1.1.1 Homogenisation techniques

The technique chosen for disrupting or homogenising any biological material depends on the nature of the sample and the nature of the product required from the extraction. Thus different techniques are required for disrupting leaves of plants, animal tissues, bacteria and for the extraction of proteins, ribonucleic acids etc. As an example, in the extraction of DNA from soft animal tissue an easily accomplished first step is brief homogenisation in a blender (Section 9.3.2). Such treatment will break many cells and will inevitably lead to some physical damage to organelles and shearing of DNA. Thus the choice of the technique is a compromise; vigorous homogenisation will produce high yields of damaged DNA while gentle homogenisation produces low yields of high quality DNA.

SAQ 1.1	List as many different techniques available for disrupting biological material as you can and indicate the type of material to which each is best suited.

1.1.2 Homogenisation media

The choice of the homogenising medium is very important; this depends to some extent on the ultimate objective, ie, whether you require the cell walls, a specific enzyme or simply the isolation of a particular type of macromolecule.

SAQ 1.2	List the main factors, with a brief explanation, which should be considered in designing a suitable homogenisation medium.

By this time you will have gathered that there is no such thing as the best or ideal homogenisation medium, each is a compromise for each particular situation.

Whatever the choice it is essential that the temperature is kept as low as possible during the disruption process.

∏ Write down a reason why it is important to maintain a low temperature.

Any of the physical or mechanical methods generate heat and unless the sample is cooled, denaturation of some macromolecules can occur, possibly resulting in the reduction of yield, loss of enzyme activity etc.

1.1.3 Contents of homogenate

∏ Before we consider any methods of separation it is worthwhile for you to spend a short time listing the different types of material or substances present in the homogenate.

The main contents of the homogenate include:

* debris from connective tissue, cell wall material;

* intact cells and subcellular fractions;

* cellular contents, macromolecules such as nucleic acids, proteins, enzymes;

* small molecules of biological origin, such as amino acids;

* components of the buffer solution, possibly with some added inorganic ions such as Mg^{2+};

* material used to maintain the osmotic pressure of the medium, such as sucrose.

1.2 Separation methods

The separation process(es) used must be selected to permit the isolation of the required material from the homogenate. There are many techniques for separation and before we consider each in more detail try SAQ 1.3.

SAQ 1.3	List the main methods available for the separation of biologically important molecules from homogenates of cellular material; for each explain the principle of the technique, for example solubility, charge.

1.2.1 Centrifugation

The first and possibly the simplest technique to use is filtration possibly through gauze or filter paper; this removes the larger debris such as connective tissues which are visible to the naked eye - this, of course, maybe the material you require for further study.

Centrifugation, in one of its various forms, is the next technique used to separate particulate matter, eg cell wall debris, from the soluble macromolecular species.

In this technique, centrifugal forces are applied on particles. The rate at which they sediment is dependent upon the size of the centrifugal force, the size of the particle and the differences in density between the particles and the suspending medium. The velocity at which particles sediment in a centrifugal field is often expressed by the following relationship:

$$v = 2\omega^2 x r^2 (\rho_p - \rho_m)\, 9\eta \qquad\qquad (E - 1.1)$$

where:

$v =$ sedimentation rate; ω is the angular velocity in radians per second (ie 2π times resolution per second).

$x =$ radial distance of the particle from the centre of the rotor

$r =$ radius of the particle

ρ_p and $\rho_m =$ densities of the particle and medium

$\eta =$ coefficient of the viscosity of the medium.

We will deal with some of the numerical aspects of centrifugation in Section 1.2.3. Here we will focus on the various types of centrifugation strategies.

The three main types of centrifugation are:

differential centrifugation
- differential centrifugation: routinely used to remove contaminating cell components, such as heavy nuclei, cell debris, cell walls, chloroplasts and mitochondria prior to density gradient centrifugation;

rate-zonal centrifugation
- rate-zonal centrifugation: routinely used for the isolation of subcellular fractions, plasma membrane and golgi apparatus from plant microsomes. In this technique centrifugation proceeds for a fixed time, in a medium of increasing density, (the density of the particles must be greater than the maximum density of the gradient medium). The particles separate in discrete zones or bands (Figure 1.1) depending on the centrifugal field, the shape and size of the particles and the relative densities of the particles and the medium.

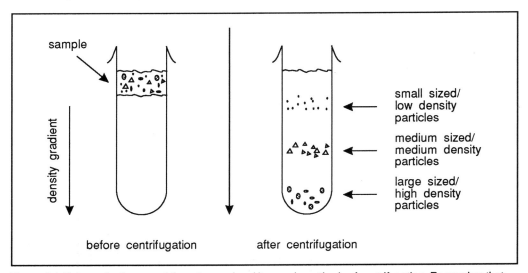

Figure 1.1 Schematic diagram of the rate-zonal and isopycnic methods of centrifugation. Remember that centrifugation separates particles because of the combined effect of size and density (see Equation (E - 1.1).

- isopynic (equal density) centrifugation; most commonly used to separate nucleic acids (Section 8.4.1). This technique, in which centrifugation occurs in a medium of increasing density, is based solely on the buoyant density of the particles, the density of the particles must never exceed that of the medium. Separation of the particles continues until their buoyant density is equal to that of the medium, after which no further sedimentation towards the bottom of the tube occurs.

In isopycnic centrifugation the continuous density gradient may be 'pre-formed' (Section 1.2.2) or 'self-formed'. Concentrated solutions of salts of heavy metals are traditionally used to generate self-forming density gradient media, caesium salts for the separation of nucleic acids and rubidium salts for the separation of proteins.

∏ Media of increasing density are used in both the rate-zonal and the isopynic technique; what do you think would happen if centrifugation was carried on for an extended period of time using either technique?

No doubt you realised that in the rate-zonal technique the density of the particles is greater than the maximum density of the gradient medium and so on prolonged centrifugation the particles will sediment. In the isopycnic technique the density of the particles never exceeds that of the medium and so on prolonged centrifugation the particles will not move from the position where their buoyant density is equal to that of the medium.

1.2.2 Density gradient medium

A density gradient medium must be carefully evaluated in terms of its chemical and physical properties.

SAQ 1.4   What are the most important parameters in the choice of an ideal gradient material?

Again you can see that there is no one ideal density gradient material, the choice is a compromise for each particular application. There are, however, certain media commercially available for the formation of gradient media, these include:

- sucrose, the most commonly used material. It is stable in solution, inert towards biological material, and inexpensive. It exhibits, however, a high osmotic pressure at low concentrations and its high viscosity at high concentrations (> 40% w/w) precludes its use in some isopycnic techniques;

- glycerol, suffers from the disadvantage of high viscosity;

- a variety of polysaccharides, eg dextrans, Ficoll (a copolymer of sucrose and epichlorohydrin). Although Ficoll exerts a low osmotic pressure it can cause acute viscosity problems;

- iodinated compounds, such as Metrizamide, Nycodenz, based on the structure of triiodobenzoic acid, exert much lower osmotic pressures and have lower viscosities than sucrose and the polysaccharides;

- colloidal silica, eg Percoll, has the advantage that its coating of polyvinylpyrrolidone prevents adherence to biological material and enhances the stability of the suspension. Only particles significantly larger than the silica particles can be separated.

The apparatus required to form a continuous density gradient consists essentially of two cylindrical chambers (A and B), connected at their base through a tap which controls the mixing of the contents of the two chambers (Figure 1.2). The mixing chamber B is stirred by a magnetic stirrer and possesses an outlet to a peristaltic pump connected to a centrifuge tube or chromatography column.

If the chambers are of equal diameter ($d_A = d_B$) and the denser solution of the chosen gradient medium is in chamber A and an equal volume of water or a less dense solution in B, then on opening the tap between the two chambers the less dense solution will gradually be replaced by one of increasing density - this experimental set up produces a linear density gradient. Linear continuous density gradients are used for the separation of nucleic acids and ribosomal nucleoproteins.

If $d_A < d_B$ then a concave exponential gradient (Figure 1.2) is formed; while if $d_A > d_B$ a convex gradient results.

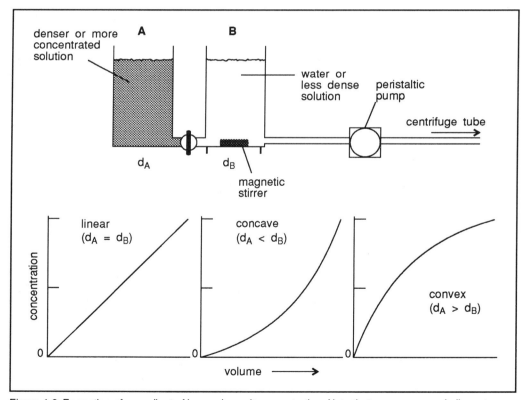

Figure 1.2 Formation of a gradient of increasing salt concentration. Note that we can use a similar set up to produce a gradient of increasing salt concentration for eluting materials from chromatography columns. We will discuss various forms of chromatography in Section 1.3.

∏ Can you see any advantages in using a concave gradient, where the initial gradient near the top of the tube is steep and the gradient near the bottom is shallow?

It can be useful in the differential separation of particles; the smaller particles near the top of the tube will separate further than the heavier particles near the bottom of the tube.

SAQ 1.5

The apparatus shown in Figure 1.2 can also be used to provide a solution of linearly decreasing pH for use in eluting ion exchange chromatography columns. Explain how this may be achieved.

1.2.3 Ultracentrifugation

The separation of macromolecules can be achieved by ultracentrifugation, using centrifugal fields in excess of 600 000 g in gradient density media; CsCl is used for the separation of DNA from contaminants and concentrated sodium acetate solutions for RNA. This technique requires costly and sophisticated instrumentation and is not widely used. There are simpler methods to achieve such separations, these will be considered in the next sections.

In the analytical mode, ultracentrifuges are used for the accurate determination of molecular mass; the results can also give an idea of the shape of the molecules. The sedimentation coefficient (constant), s, a parameter often used to describe the size of proteins and nucleic acids, is defined as:

$$s = (1/\omega^2 x) (dx/dt) \tag{E - 1.2}$$

where ω is the angular velocity of the rotor (ie 2π x revolutions per second), x the radial distance of the particle from the centre of the rotor and dx/dt the rate of sedimentation. If the particle (boundary) moves from a position x_1 at time t_1 to a position x_2, at time t_2, Equation (E - 1.2) can be integrated to give:

$$s = (\ln x_2/x_1)/[\omega^2(t_2 - t_1)] \tag{E - 1.3}$$

sedimentation coefficient

The sedimentation coefficient (units:second) can be determined by measuring the position of the boundary of the particles (molecules) at fixed intervals. The sedimentation coefficient is constant for a given solute molecule in the medium used. For proteins the value of s falls in the range 10^{-13} to 200×10^{-13} s, the unit 10^{-13} s is called a Svedberg, (recommended symbol: Sv, although most books use the symbol S. Here we will use the Biologist's convention of using S) named after the worker, Svedberg, who initiated the development of the ultracentrifuge in 1923. Proteins and nucleic acids are often classified in terms of their sedimentation coefficients, eg 20 S, 40 S. For more details of the use of analytical centrifuges and the mathematical treatment you are advised to consult the BIOTOL text, 'Techniques used in Bioproduct Analysis'.

| **SAQ 1.6** | The sedimentation of bovine serum albumin was monitored at 25°C. The initial distance of the solute surface from the centre of the rotor was 5.50 cm and during centrifugation at 56 850 rpm it receded as follows: |

time(s)	0	500	1000	2000	3000	4000	5000
x(cm)	5.50	5.55	5.60	5.70	5.80	5.91	6.01

Calculate the value of the sedimentation coefficient.

| **SAQ 1.7** | Indicate whether each of the following statements is true or false. |

1) The force on a particle in an applied centrifugal field depends only on its distance from the centre of rotation.

2) Sedimentation of a heterogeneous suspension of particles results in a sediment which is homogeneous.

3) The rate of sedimentation of a macromolecule depends on its shape as well as its molar mass.

4) The way a sample is loaded is critical in isopycnic centrifugation.

1.3 Chromatographic techniques

1.3.1 Introduction

The separation of molecules from biological materials is an important part of biochemical work and often involves the isolation of one particular molecular species from a mixture of compounds with similar properties.

∏ Explain why the usual methods of organic chemistry are inadequate.

To prevent loss of biological activity resulting from denaturation, the following should be avoided:

• extremes of temperature and pH;

• the use of organic solvents;

• the use of oxidising and reducing agents;

The experimental techniques used for such separations must be carried out under mild conditions and make use of small differences in the physical properties of the molecules.

∏ List some physical properties of molecules which could be employed to separate molecules.

Your list should have included:

- such basic properties as mass, charge, shape and size;

- solubility;

- adsorptive properties;

The methods to be described generally involve differing degrees of interaction between the mixture to be separated, a solid phase and a solvent. The magnitude of the interaction depends on the particular technique used; thus in ion exchange chromatography solute-solid interaction is dominant, while in partition chromatography it is solute-solvent. There are a wide range of chromatographic techniques each with its own particular advantages in a given situation. We will now briefly review the principles and practice of the more important of these techniques.

1.3.2 General principles of chromatography

The International Union of Pure and Applied Chemistry (IUPAC) defines chromatography as, 'A method used primarily for the separation of components of a sample, in which the components are distributed between two phases, one of which is stationary while the other moves. The stationary phase may be a solid, or a liquid supported on a solid, or a gel. The stationary phase may be packed in a column, spread as a layer or distributed as a film, etc. In these definitions; 'chromatographic bed' is used as a general term to denote any of the different forms in which the stationary phase may be used. The mobile phase may be gaseous or liquid.'

normal phase chrom- atography
If the stationary phase is more polar than the mobile phase the technique is referred to as normal phase chromatography and if the stationary phase is less polar than the mobile phase the technique is called reverse phase chromatography.

In practice almost all forms of chromatography are carried out with the stationary phase in the form of fine particles (to provide a large surface area) packed into a column over which is passed the mobile phase. A small volume of the sample is loaded on the top of the column and the mobile phase carries the mixture down the column. As it does so, the solutes are in constant dynamic equilibrium between the stationary and mobile phases. Separation is the result of differences in the way in which the solute molecules distribute themselves between the two phases.

If a molecule A interacts more strongly with the stationary phase than does a molecule B, then A will have a longer retention time than B. After passing through a detecting system, which may be electrical or optical, the effluent is collected as discrete volumes in a fraction collector. Analysis of the fractions for a particular compound(s) provides an elution profile, ie quantity of material eluted against effluent volume or retention time (see Figure 1.5).

high performance liquid chrom- atography
High performance liquid chromatography (HPLC) differs from classical chromatography mainly in the nature of the particles of the stationary phase. This consists of very small (5 μm diameter) regularly sized rigid particles which provides a very high surface area. A high pressure (up to 600 atmos) is required to force the mobile

phase through the densely packed, narrow bore (1-7 mm internal diameter) metal column. This technique has enormous advantages over gravity-fed columns in terms of efficiency, speed of separation, and reproducibility; furthermore it is applicable to all forms of chromatography.

1.3.3 General chromatographic techniques

adsorption chrom-atography

Adsorption chromatography is the simplest of all techniques; separation occurs because different molecules adsorb and are displaced differently. Because adsorption is a purely physical process, a wide range of materials is available as the stationary phase, for example silica, alumina, cellulose and carbon.

SAQ 1.8	Which of the following types of interaction are most likely to be involved in the process of adsorption; 1) ionic interaction; 2) dipole-dipole interaction; 3) van der Waals forces; 4) hydrogen bonds?

If the stationary phase is polar (eg silica, alumina), polar molecules will adsorb strongly, whereas non-polar molecules will be less strongly adsorbed. As the solvent displaces the solute by competition for the adsorption sites, the effectiveness of a solvent depends on the strength of the interaction with the adsorbent, thus for alumina the elutropic series is water > ethanol > propanol > trichloromethane > benzene > pentane. By the term elutropic we mean 'ability to elute'.

partition chrom-atography

Partition chromatography is intermediate between adsorption chromatography and ion-exchange chromatography; compounds soluble (but to a different extent) in both aqueous and organic solvents can be readily separated by partition methods. When two immiscible liquids are shaken, a compound distributes itself unevenly between the two phases A and B, and at equilibrium the partition coefficient K_D is defined as:

$$K_D = \frac{\text{concn of solute in phase A}}{\text{concn of solute in phase B}} \qquad \text{(E - 1.4)}$$

or for a water/organic solvent system:

$$K_D = c_o / c_{aq} \qquad \text{(E - 1.5)}$$

In partition chromatography one solvent, usually water, is held on the stationary support medium. The support medium may be either in the form of a column or as a thin film of inert material. The mobile phase is usually a water saturated organic liquid. The components of the mixture can be separated if their partition coefficients are sufficiently different.

Cellulose, in the form of paper, makes an ideal support; here the water is adsorbed between the fibres and forms the stationary hydrophilic phase. Development occurs when the organic phase flows over the paper (usually by ascending up the paper). The paper is dried and the components visualised using specific reagents. The R_f value is defined as:

$$R_f = \frac{\text{distance moved by compound}}{\text{distance moved by solvent front}}$$

and is characteristic of the compound in a given solvent system. Separation can also be carried out in two dimensions; after separation in solvent 1 and drying, the paper is turned through 90° and developed in solvent 2 (Figure 1.3). This gives enhanced separation of components in complex mixtures.

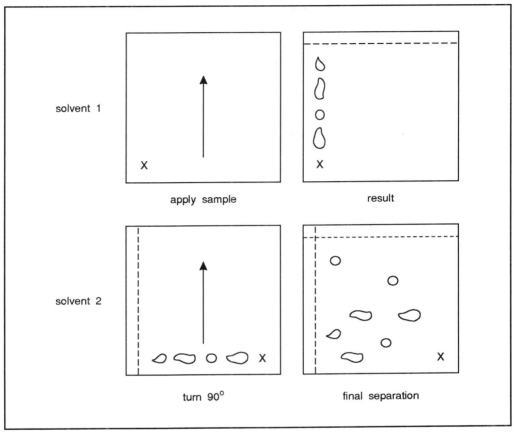

Figure 1.3 Separation of a mixture by two-dimensional chromatography. (Redrawn from Plummer D.T. 1978, ' An Introduction to Practical Biochemistry', 3rd Edition, McGraw Hill).

reverse phase chrom-atography

Reverse phase chromatography is a modified form of partition chromatography in which the mobile phase is hydrophilic and the stationary phase hydrophobic. Columns (often in the form of HPLC) and TLC plates consist of silica particles, to which are bonded alkylsilyl groups (2 to 22 carbons in length), are used in conjunction with the more mobile polar phase commonly a mixture of water with methanol, ethanol or ethanenitrile (acetonitrile). Ion-pairing of peptides, for example with trifluoroacetic acid, increases their hydrophobility and hence enhances their separation by RPC (Section 4.2.1).

∏ What do you think is the mechanism for retention in reverse phase chromatography?

In this type of chromatography retention occurs by hydrophobic interaction between the solute and the stationary phase; the wide application of this type of chromatography is due to the fact that most bio-molecules have hydrophobic regions in their structure which are capable of interacting with the stationary phase.

∏ What type of molecules would you expect to have the longest retention time on a reverse phase column?

Polar solutes will move with the polar mobile phase, while non-polar solutes will be retained by the non-polar stationary phase and hence have longer retention times.

It is important to recognise the difference between normal and reverse phase chromatography since the approach to changing the solvent depends on the system we are using. Remember the principle 'like dissolves like' in answering the following SAQ.

SAQ 1.9

Complete the following table using the words provided below:

Type of chrom- atography	Polarity of stationary phase	Polarity range of mobile phase	Order of elution polar/ non-polar	Effect of increase of polarity of mobile phase
normal phase				
reverse phase				

non-polar; polar; weak to medium polarity; strong to medium polarity; most polar eluted first; most polar eluted last; retention times decreased; retention times increased.

Reverse phase chromatography is used in the separation of peptides (Section 4.2.1), and derivatised amino acids (Section 5.3.2). Silica impregnated paper is available commercially for the separation of lipids and similar hydrophobic molecules. HPLC is used extensively for the separation of peptides (Section 4.2.1).

gas-liquid chrom- atography

Gas-liquid chromatography (GLC) is another partitioning technique, this time between a liquid held on a stationary phase and an inert gas (eg nitrogen, argon, helium). The liquid stationary phase (eg polyethylene glycols, silicone gums) is coated on an inert support in a long narrow coiled tube (1-3 m long, 1-4 mm internal diameter), the sample is injected as a vapour at the head of the column. The rate at which the components are eluted depends on their partition coefficients. Qualitative identification of the compound is based on the time (retention time) for it to appear in the detector; quantitative data are obtained from the evaluation of the peak areas on the chromatogram. GLC, an exclusively analytical tool for very small amounts of sample, is used to separate non-polar, volatile compounds (eg hydrocarbons, esters) and has, to some extent been replaced by reverse phase HPLC.

thin layer chrom- atography

Thin layer chromatography (TLC) is a technique in which separation is carried out on a layer of supporting medium spread in a thin film (150 to 250 μm thick) on a glass, aluminium or plastic sheet. Separation can be by adsorption, partition, ion-exchange or gel filtration depending on the nature of the support medium. Most chromatograms are run in the ascending form and when dry the various components are located by physical or chemical means. Like paper chromatography, TLC can be carried out in two dimensions and is also used as a second dimension after electrophoresis.

Π What are the advantages of carrying out two-dimensional chromatography?

For a given solvent system each compound has a specific R_f value. If there are two compounds with very similar R_f values in one solvent system they are unlikely to have the same R_f values in a second solvent system, and so they will be separated during development in the second dimension. For complex mixtures, a single dimension chromatogram may well be difficult to interpret due to close or overlapping spots. Development in a second dimension will resolve the spots and make for easier interpretation.

1.3.4 Specific techniques

ion-exchange chrom-atography

Ion-exchange chromatography is the separation of charged species on an insoluble matrix containing lattice ions which are capable of exchanging with ions in the surrounding medium. Mixtures are passed through a column containing an ion-exchange resin (cross-linked polymers carrying acidic or basic ionisable groups), if the groups are negatively charged (cation exchanger) they will attract positively charged species from the mixture. Positively charged (anionic) exchangers attract negatively charged species from the mixture.

The exchange process occurs by displacement of counter ions of the exchanger by the ions in the sample. The bound ions can be eluted by changing the ionic strength or the pH of the mobile phase, usually in the form of a gradient solution. A typical separation on and regeneration of an anionic exchange resin is shown schematically in Figure 1.4.

SAQ 1.10

You have an ion-exchange column on which a mixture of proteins is bound. What happens when the pH of the eluting medium is progressively changed?

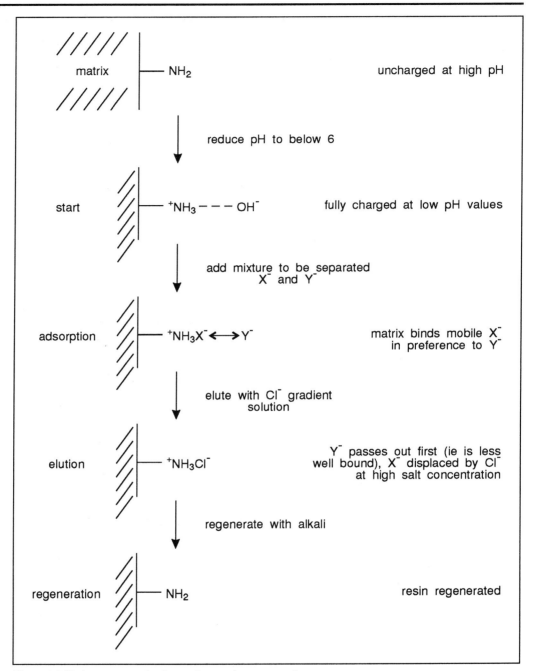

Figure 1.4 Various stages of anion exchange chromatography.

Ion-exchange chromatography (in the form of a column, on paper or TLC) is easy to carry out and is a powerful technique for separating closely related molecular species, eg proteins (Section 3.3.6) amino acids (Figure 1.5) and derivatised amino acids (Section 5.2.3). We will discuss the separation of these components more thoroughly in a later chapter.

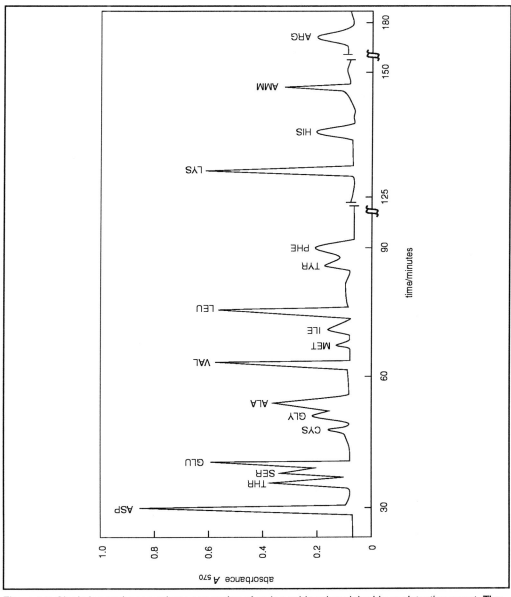

Figure 1.5 Single ion exchange column separation of amino acids using ninhydrin as detecting agent. The resin was used in an HPLC column (pressure 800 psi). Mobile phase sodium citrate buffer solutions (pH 3.27 to pH 5.9). Note that the abbreviations used for each amino acid are described in a table in an appendix.

gel filtration
chrom-
atography

Gel filtration chromatography or gel permeation chromatography (GPC) is a separation technique dependent on the shape and size of the solute molecules. The mixture, applied at the top of a column of the gel, is washed through with water, buffer solution or organic solvent. Solutes with molecules larger than the largest pores of the swollen beads (ie above the exclusion limit) cannot penetrate the particles and in consequence pass through in the liquid phase and elute first. Smaller molecules penetrate the pores to varying extents, depending on their size and shape, this results in partitioning between the liquid inside the gel and that outside (Figure 1.6).

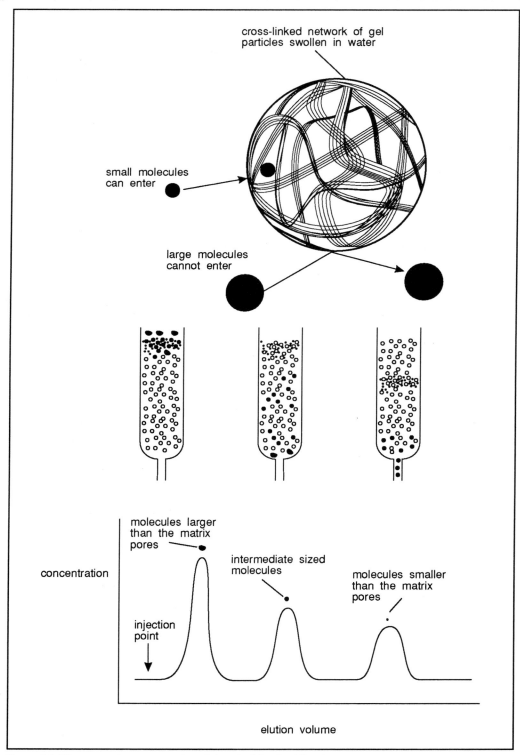

Figure 1.6 Diagrammatic representation of the principle of gel filtration.

The gels are based on cross-linked dextrans, polyacrylamides and high molecular mass polysaccharides (eg agarose); the amount of cross-linking determines the swelling properties of the gel and its exclusion limits, and hence the range of size of molecules that can be separated. The technique, involving no chemical reactions, is suitable for the separation of biological macromolecules, such as proteins (Section 3.3.5), nucleic acids (Sections 7.3.4 and 10.4.2) and is commonly used for desalting solutions. Gel filtration chromatography may also be used for approximate determination of the relative molecular mass (RMM), the elution volume is approximately a linear function of log(RMM).

∏ Is it possible to separate two proteins A and B of the same RMM, in which A is a tightly coiled spherical molecule while B is a long rod shaped molecule?

Yes it is, A being spherical is more compact and is effectively a smaller molecule than B; therefore it may penetrate the gel beads to a greater extent than molecule B. Molecule B, on the other hand, occupies a much larger three-dimensional volume (due to its continual random rotation) as it will not be able to penetrate the pores of the gel. Molecule B will, therefore, migrate through the column first.

affinity chrom-
atography

Affinity chromatography is an elegant technique for isolating biological molecules which depends on the specific affinity of one molecule for another. The biospecific adsorbent, ie the stationary phase is prepared by coupling a specific ligand (for example enzyme, antigen, hormone) for the macromolecule of interest, to a gel matrix (such as agarose) usually via a spacer arm.

∏ Try to explain why a spacer arm is used.

It is necessary to hold the ligand away from the surface of the supporting medium to avoid steric hindrance thus allowing access for large molecules to the ligand.

The mixture to be separated is passed down the column and those molecules with specific binding affinity for the ligand are retained whilst the rest of the mixture remains unbound and is eluted from the column (Figure 1.7). The bound molecules can be eluted from the column by altering the pH, ionic strength or by displacement with molecules with greater affinity for the ligand. For the separation of proteins, often antibodies are used as the immobilised binding agent. Use is made of the antibody's specificity to bind specific protein molecules to the column. Affinity chromatography offers many advantages compared to conventional methods for the separation of proteins (Chapter 3) and RNA (Chapter 11): selective adsorption can increase the purity by 1000-fold, the number of purification steps is reduced, the species is concentrated and frequently stabilised on binding, and the method is suitable for large scale purification processes.

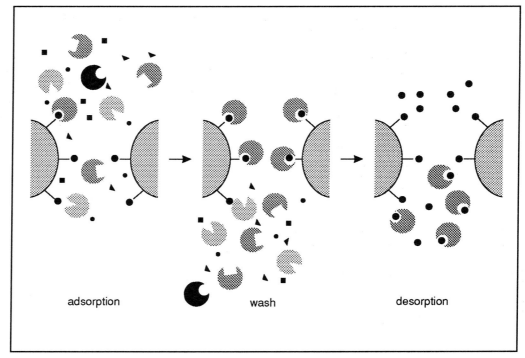

Figure 1.7 Principles of affinity chromatography (see text for explanation).

1.3.5 Summary of chromatographic methods

Characteristic	Separation method	Resolution	Speed	Reproducibility
partition	partition chromatography	good	low	good
	gas/liquid chromatography	very good	fast	very good
size	gel filtration	good	low	very good
charge	ion exchange	very good	fast with correct matrix	very good
hydrophobic nature	reverse phase	very good	fast	very good
biological activity	affinity chromatography	excellent	fast	excellent

Note that polyacrylamide gel electrophoresis is sometimes regarded as a chromatographic technique. Here however we will consider it under electrical methods.

SAQ 1.11	Complete the following table which gives a summary of the more important types of chromatography.

Name of technique	Nature of stationary phase	Mobile phase	Mechanism of separation	Applications, separation of:
gas liquid chromatography				volatile esters, hydrocarbons
adsorption chromatography				
reverse phase chromatography				
partition chromatography			partition liquid/liquid	
ion exchange chromatography				
gel filtration chromatography	cross-linked gel (eg agarose)			
affinity chromatography				antigens

1.4 Other separation techniques

1.4.1. Electrical methods

electrophoresis

Electrophoresis. Many biological molecules carry an electrical charge, the magnitude of which depends on the particular molecule, the pH and composition of the suspension medium. These charged species migrate in solution to the electrode of opposite charge when an electric field is applied. This is the principle of electrophoresis for the separation of differently charged species.

The charge carried is proportional to the electrophoretic mobility, u, defined as the rate of migration (v ms^{-1}) per unit applied field strength (a V m^{-1}), ie $u = v/a$. The mobility is the quantity which can be measured experimentally. In moving boundary electrophoresis, a technique first used by Tiselius, the migration of the species under an applied field is monitored by changes in the refractive index of the solution. The method has many disadvantages; it requires very careful control of temperature during the course of an experiment which may last for 24 hours; it is subject to convection currents which disturb the boundaries, it can be used to determine the mobility of separate species but not to separate them and it is time consuming.

zone electrophoresis

To overcome these problems electrophoresis is now carried out on a supporting medium impregnated with buffer solution. Complete separation of a mixture can be effected into discrete zones, hence the name 'zone electrophoresis'. Filter paper, a cheap support medium, suffers from the disadvantage that some mixing occurs between the zones due to adsorption onto the cellulose. It has been replaced by cellulose acetate which shows minimal adsorption and gives clear separation into discrete zones. Agar is often the choice for immunoelectrophoresis. Note that separation in agarose and polyacrylamide gels is by size and not by charge. We will deal with this later.

The basic components of an electrophoresis apparatus are shown in Figure 1.8. The support medium is saturated with a buffer solution of appropriate pH, the sample applied as a spot or band near one end, and a stable electric field (2 to 10 V cm⁻¹) applied. When electrophoresis is complete the medium is dried and the zones visualised.

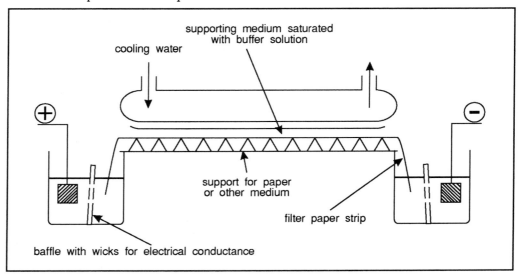

Figure 1.8 The basic components of an electrophoresis apparatus. Redrawn from Hibbert, D.B. and James, A .M. 1984, 'Dictionary of Electrochemistry,' Macmillan, London.

∏ Can you explain why in the apparatus depicted in Figure 1.8 it is necessary to cool the medium and why the electrode compartments are separated from the support medium by wicks?

The passage of an electric current through a thin film of buffer solution generates a considerable amount of heat, this in turn would cause evaporation of water from the film resulting in convection currents and changes in the ionic strength of the buffer solution. The electrode compartments are separated from the support medium by wicks of filter paper to minimise products of hydrolysis (H^+ or OH^-) reaching the buffer solution thereby causing a change in pH. Increased separation of a complex mixture can be achieved by zone electrophoresis in one direction followed by conventional TLC in a second direction.

gel electrophoresis **Polyacrylamide gel electrophoresis,** (PAGE) is a technique in which migration of the solutes, in a gel, occurs as a result of their charge while separation is a consequence of their size or a combination of size and charge. The gels are prepared by cross-linking the acrylamide monomer with a small amount of N,N' methylenebisacrylamide in the presence of an initiator; the size of the pores depends on the extent of cross-linking which is controlled by the relative amounts of the monomer and bisacrylamide. The gels are usually prepared as 'slab' gels in which there are a series of wells to load the samples under investigation. Buffer solutions are incorporated to ensure that all the components carry the same charge and so migrate in the same direction. The whole slab is mounted in an electrophoresis apparatus (see Figure 1.8) and a potential gradient applied; migration is opposed by the pores of the gel which act as a molecule sieve, smaller species migrate faster and further through the gel, while larger species are hindered. When separation is complete the gel is stained to visualise the components.

The technique is extensively used in the separation of proteins (Section 2.5) and nucleic acids. We will meet with this many times in subsequent chapters.

A special form of polyacrylamide electrophoresis is sodium dodecyl sulphate polyacrylamide gel electrophoresis (SDS-PAGE). In this technique, a protein mixture is first denatured with SDS and 2-mercaptoethanol which results in a reduction of the S-S-bridge in the protein and a dissociation of the polypeptide chains. The SDS forms a complex with the polypeptides. These complexes have a strong negative charge that completely overshadows the charge of the polypeptide itself. The SDS-polypeptide complex carry an almost uniform charge density and also the form of the complexes is rather regular. Therefore, the migration velocity through the gel is determined by the molecular mass of the polypeptide. We will be applying this technique to the determination of the relative molecular mass (RMM) of polypeptides in Chapter 2.

Polyacrylamide gel electrophoresis can also be used to determine the relative molecular mass of nucleic acids (even with short chains of nucleotides). In this case, we do not have to use SDS as nucleic acids have more-or-less constant negative charge densities per unit length of their nucleotide chain. For large nucleic acid molecules, however, the pores in polyacrylamide gels are too small for the nucleic acid molecules to penetrate. In this case a more open gel (usually agarose - a gel forming polysaccharide) is used.

Immunoelectrophoresis. No discussion of electrophoresis would be complete without mention of immunoelectrophoresis. It is especially valuable when we need to detect very small amounts of protein. In this technique, the sample is first separated by electrophoresis, then antibodies are allowed to diffuse into the gel. A precipitation reaction occurs when an antibody reacts with its appropriate antigen providing the two are present in approximately equal proportions.

We can represent this in the following way:

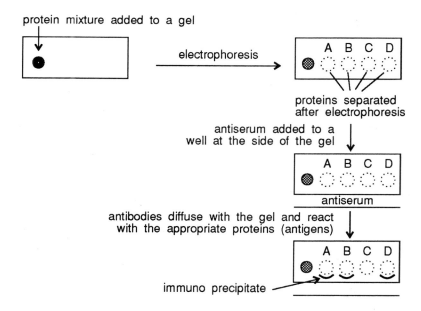

∏ In the sequence shown above, for which protein is there no equivalent antibody in the antiserum?

You should have spotted that it was C as there was no immunoprecipitate for this protein.

Laurell rocket electrophoresis, is a technique where the gel in which the electrophoresis takes place contains an antiserum against an antigen present in the sample. The samples are brought into the sample wells. After application of the electric field, the antigen will move through the gel, forming a complex with the antibody molecules it finds on its way. The complex formation will continue until the equivalence point (when antigen and antibody are present in roughly equal amounts) is reached and precipitation of the complex occurs. The result is a precipitation line with the form of a rocket, the surface of which is equivalent with the amount of antigen in the sample.

Blot techniques. The principle of the technique is simple: DNA, RNA or protein mixtures are separated by gel electrophoresis, extracted from the gel and transported to a membrane where they are bound and can be detected. The extraction from the gel can proceed via diffusion, in which case a membrane is used at both sides of the gel; via transport in a solvent that moves through the gel from a solvent reservoir to a pile of chromatography paper or via electrophoresis (see Figure 1.9). In extraction by solvent and extraction by electrophoresis one blot per gel made. Electrophoretic blotting is the fastest procedure and therefore most often used.

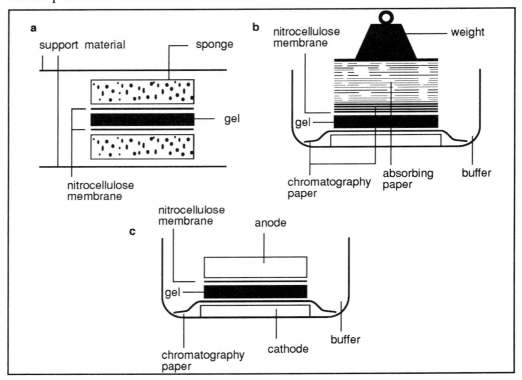

Figure 1.9 Transport of a protein to a blot membrane by diffusion (a), transport in a solvent (b), and electrophoresis (c).

The procedure which is used to detect specific DNA sequences is called *Southern blot*, specific RNA sequences are detected in *Northern blot* and proteins in *Western blot*.

We will discuss the detection of specific nucleotide sequences (Southern and Northern blotting) in Chapter 6. In Western blots, the presence of particular proteins is usually detected using antibodies.

isoelectric focusing — **Isoelectric focusing** separates ampholytes (eg proteins) in a pH gradient under an applied electric field. Molecules possessing a net charge migrate in an electric field towards the region in which they are isoelectric. At the isoelectric point, the molecules have no net charge and therefore do not migrate in the electric field; they therefore concentrate in a narrow zone. Separation of components is solely in respect of one parameter; the isoelectric point. The principle of separation of two proteins by this method is shown in Figure 1.10.

Consider molecule B. If it is anywhere in the column at a pH below 8 (its isoelectric point), it will carry a net positive charge. Within the electric field it will begin to migrate towards the negative electrode. In so doing, it will pass through regions of higher pH until it reaches pH 8. At pH 8, it will carry no net charge and will cease to migrate.

Carrier ampholytes (such as polyaminopolycarboxylic acids, with a range of pI values) required to form the pH gradient must have good buffering capacity and solubility in water at their isoelectric points, pI. These carrier ampholytes can be incorporated in polyacrylamide and agarose gels. The apparatus used for electrofocusing is similar to that used for zone electrophoresis (Figure 1.8). Although the method is capable of handling quite large sample loads, smaller loads lead to higher resolution. Once separation is achieved, the field is switched off and the zones revealed by staining with dyes, silver salts or by immunofixation.

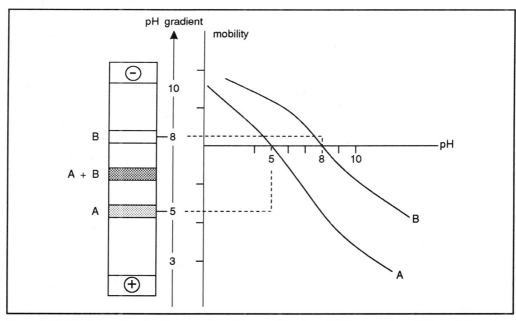

Figure 1.10 Principles of isoelectric separation of two proteins A and B. Redrawn from Hibbert, D.B. and James, A.M. 1984, 'Dictionary of Electrochemistry'. Macmillan, London.

SAQ 1.12

In Figure 1.10 the mixture of proteins A + B is applied at pH 6 (the hatched area) in a linear pH gradient. Explain how the proteins are separated as shown in the diagram.

1.4.2 Other methods of separation

solvent extraction

Solvent extraction is a simple technique based on the solubility properties of a molecule in two immiscible liquids (see Equations (E - 1.4) and (E - 1.5)). A non-polar molecule in an water/organic solvent system will have a high K_D because it is likely to be more soluble in the organic phase, whereas a polar molecule is likely to partition mostly into the aqueous phase. In this way a compound can be selectively extracted into one of the two phases.

∏ If the molecule of interest is insoluble in an organic solvent, can you explain why extraction with an organic solvent may be advantageous?

Extraction might remove from solution compounds soluble in organic solvents (eg lipids, fatty acids), thereby purifying the mixture.

Solvent extraction is a simple technique carried out by mixing the two phases in a separating funnel, allowing them to separate out and draining off the more dense solvent. Repeated extractions of either or both layers will give rise to improved separation.

Counter current distribution is essentially a more elaborate form of liquid-liquid extraction in which a fresh sample of one or both phases is added after each distribution. It can handle relatively large quantities of material (eg 20 g), and since it is a multistep process can separate closely related compounds with similar K_D values.

precipitation **Precipitation** is a crude purification process in which the dissolution of relatively large amounts of a compound (usually an inorganic salt) causes the less soluble components of a mixture to precipitate. This procedure is also known as 'salting out'. When the ammonium sulphate concentration of a crude mixture of proteins is slowly increased so more and more protein precipitates out of solution, this enables a crude fractionation of the protein mixture. The proteins are not denatured, they can be recovered by centrifugation and dissolving the precipitates in buffer solution (Section 3.3.3).

∏ Give a qualitative explanation of this process.

Proteins in solution are stabilised with water molecules. The added ammonium sulphate, especially at very high concentrations, competes for these water molecules thereby exposing the hydrophobic regions on the protein. The proteins then form insoluble aggregates by hydrophobic interaction, thus protein-protein interaction becomes more important than protein-water interaction.

Precipitation can also be brought about by the addition of organic solvents, these lower the relative conductivity of the medium and also compete for the water molecules on the protein. This method must be carried out at very low temperatures to avoid denaturation of the protein.

1.5 Spectroscopic techniques

In this final section of the chapter we will briefly revise some analytical tools that have been applied to the analysis of proteins and nucleic acids. If you are unfamiliar with these techniques we would recommend you read the relevant sections of, 'Techniques used in Bioproduct Analysis', or some other good analytical techniques text. Many chemicals adsorb electromagnetic radiation. The adsorption of such radiation is dictated by the amount and the structure of the chemical and on the environment in which it is found. By measuring the adsorption characteristics, we can therefore gain much valuable information. The most common forms of sepctroscopy employed in the study of biomolecules uses light of the visible and ultra violet (UV) ranges of the spectrum. We can use such spectroscopy in many ways. For example, adsorption characteristics can be used to determine the amount of particular chemicals, as a check for purity, to monitor elution profile from chromatographic columns and for monitoring the progress of reactions. Spectroscopy however is not confirned to the use of UV and visable wavelengths. The adsorption of longer wavelengths (infrared) gives valuable structural information. In this section we will briefly review the spectroscopic techniqes which have found widespread use in the analysis of bio-molecules.

1.5.1 Introduction

The wavelength, λ, ie the distance between two adjacent peaks (maximal wave amptitudes), and the frequency, ν, ie the number of wave cycles per second (units s^{-1}, Hz) are related by the expression:

$$\lambda = c/\nu \qquad\qquad (E - 1.6)$$

where c is the speed of light ($c = 2.998 \times 10^8 m\ s^{-1}$).

In infrared spectroscopy the term wavenumber, $\bar{v}$, is used, this is the reciprocal of the wavelength and the units are usually cm^{-1}.

$$\bar{v} = 1/\lambda = \nu/c \qquad\qquad (E - 1.7)$$

Electromagnetic radiation can be considered not only as a wave but also as a stream of discrete particles of energy - quanta or photons. The energy, E, of a quantum is related to its frequency by:

$$E = h\nu \qquad\qquad (E - 1.8)$$

for 1 mole this becomes:

$$E_M = N_A h\nu \qquad\qquad (E - 1.9)$$

where h is Plank's constant ($h = 6.625 \times 10^{-34}$ J s), and N_A the Avogadro constant ($N_A = 6.022 \times 10^{23}$ mol^{-1}).

SAQ 1.13	State which of the following statements are true or false:

1) the energy of a quantum is directly proportional to the wavelength;

2) the energy of a quantum is directly proportional to its frequency;

3) the frequency of ultraviolet radiation is less than that of infra red;

4) the energy of infrared radiation is less than that of ultraviolet.

If radiation of intensity I_o is incident on a sample and I is the intensity of radiation passing through the sample, then the absorbance, A, which is a measure of the radiation absorbed by the sample is given by:

$$A = \log(I_o/I) \qquad\qquad (E - 1.10)$$

Spectrophotometers for use in the UV and visible region of the spectrum normally give a record or trace of absorbance at varying wavelengths. Infrared spectrometers, on the other hand, normally record the transmittance (expressed as a percentage) at varying wavenumbers. The transmittance, T, is defined by:

$$T = (I / I_o) \qquad \text{(E - 1.11)}$$

∏ What is the relationship between absorbance and transmittance?

If Equations (E - 1.10) and (E - 1.11) are combined you obtain the relationship:

$$A = \log(I/T) = -\log T \qquad \text{(E - 1.12)}$$

Beer-Lambert law The Beer-Lambert law relates the measured absorbance to the concentration, c, and the path length, l, of the sample:

$$A = \varepsilon c l \qquad \text{(E - 1.13)}$$

where ε is the molar extinction coefficient. This law is a limiting law and is only valid over a limited range of concentrations. This relationship is used extensively in biochemistry and biotechnology to determine the concentration of compounds in solution. For this purpose it is essential to plot a calibration curve for known concentrations of the compound and ensure that it is linear and passes through the origin; since there may be deviations from ideal behaviour at higher concentrations extrapolation of the plot should not be attempted.

1.5.2 Ultraviolet and visible spectroscopy

To see if you understand the basic principles of spectroscopy attempt SAQ 1.14 before proceeding.

SAQ 1.14

The ultraviolet absorption spectra of molecules are due to which one of the following:

1) rotation of the molecules;

2) vibration of atoms within the molecules;

3) electronic transitions from the ground state, or from low energy levels to high energy levels;

4) electronic transitions from high energy levels to low energy levels or to the ground state.

Ultraviolet and visible spectra are not generally useful in the determination of the absolute structure of compounds of biological materials because the peaks are too broad. They can be used in the identification of compounds or classes of compounds. Peptide bonds absorb at about 220 nm (Section 2.2.2), while proteins also show an absorption maximum at about 280 nm (Section 2.2.2), and nucleic acids at about 260 nm (Section 7.3).

∏ What part of a protein molecule do you think gives rise to absorption at 280 nm?

It is the amino acids with side chains containing aromatic (and hence conjugated) moieties - phenylalanine, tryptophan and tyrosine - that give rise to this absorption.

1.5.3 Infrared spectroscopy

When molecules absorb infrared radiation, the energy absorbed is insufficient to cause excitation of the electrons but it can cause atoms to vibrate about the covalent bonds that bind them. Not all molecular vibrations result in the absorption of energy; for this to happen the dipole moment of the molecule must change as the vibration occurs. The intensity of the infrared absorption is proportional to the amount by which the dipole moment changes. Thus highly polar groups tend to produce very intense infrared absorptions whereas symmetrical vibrations, such as around the C=C bond of ethene, do not give infrared absorptions.

∏   What factors do you think determine the frequency of vibration?

There are two main factors which determine the frequency of a given vibration: the stiffness of the bond and the masses of the bonded atoms. Thus double bonds vibrate at higher frequencies than do single bonds and heavy atoms vibrate at lower frequencies than do light atoms.

Infrared spectra are widely used in the identification and structural study of organic molecules; such spectra are very complex. For this reason infrared spectroscopy is not usually applied to large biological molecules (eg proteins, nucleic acids, polysaccharides) but is used to identify or characterise small purified compounds (such as drug metabolites).

1.5.4 Nuclear magnetic resonance spectroscopy, NMR

Here we do not intend to describe NMR in detail. A fuller description is given in the BIOTOL text, 'Techniques used in Bioproduct Analysis', and further references are given in the, 'Suggestions for Further Reading', section provided at the end of the text. We do, however, briefly remind you of the principles involved.

Nuclear energy levels of elements having nuclear spin (eg ^{1}H, ^{12}C, ^{15}N) are split by a magnetic field. Transitions between these levels may be induced by a radio frequency. A plot of the intensity of absorption against radio frequency (at fixed magnetic field) is the NMR spectrum. The position of lines characteristic to a nucleus in a given environment (chemical shift) aids identification of the molecular structure, and the integrated intensities give the relative numbers of the particular nuclei.

Although NMR can be used for the study of the structure of a protein the spectrum is very complex and requires computer analysis as well as the use of high resolution instruments. The NMR spectrum of a relatively simple molecule such as lysine is already quite complex. NMR is more likely to be used in the detection of small conformational changes, for example in binding studies at an active site of an enzyme.

Now that we have briefly revised the basis of the important separation and analytical technique, we can turn our attention to the purification and analysis of the amino acids, proteins and nucleic acids.

Summary and objectives

Physical and chemical techniques of cell disruption are discussed with reference to their particular application. The function of the different components of the homogenisation medium are discussed.

Particles can be separated according to their size, shape and density by one of the centrifugation techniques:differential, rate-zonal and isopycnic centrifugation. The formation and use of density gradient media is described: there is no ideal medium, each is a compromise dependent on the system under study.

The properties used in the separation of biological materials include solubility, size, density, shape, charge and adsorption. Many of these properties are employed in the various forms of chromatography in which compounds partition themselves in a characteristic way between two phases. Some of the techniques are more suited to analytical purposes rather than to the recovery of the purified compound. The relative advantages and applications of the various methods are discussed.

Ultraviolet and visible spectra are the result of the absorption of energy which gives rise to the transition of electrons from a lower to a higher energy level while infrared spectra arise from the absorption of energy which causes atoms within a molecule to vibrate.

The main parameters used in spectroscopy are defined and the Beer-Lambert relationship is defined and its quantitative application considered.

On completion of this chapter you should now be able to:

- describe the main methods of cell disruption and select a particular procedure to disrupt particular cells or tissues;

- design a suitable homogenisation medium;

- explain the principles of differential, rate-zonal and isopycnic centrifugation;

- describe the formation of discontinuous and continuous gradient media; and explain the choice of materials for the required medium;

- explain the principles involved in the choice of a separation method for a particular mixture;

- classify chromatographic techniques and describe the basic principles underlying the various forms of chromatography;

- explain the principles and give a description of the different electrophoretic and blotting techniques;

- define the terms energy, frequency, wavelength and wave numbers and state their interrelationships;

- explain the reason for electronic spectra in the UV region and vibrational spectra in the infrared region;

- state the Beer-Lambert relationship and explain its use in quantitative measurements.

Detection and estimation of peptides and proteins

Detection and estimation of peptides and proteins

2.1 Introduction

separation of
proteins

During the course of protein or peptide purification it is essential that we have a method for monitoring for the presence of protein or peptides in solution. For example, in ion exchange chromatography (see Section 1.3.4 or BIOTOL text, 'Techniques used in Bioproduct Analysis',) we would load a complex mixture of proteins onto the column and elute the proteins off by applying a salt gradient. In practice, therefore, we would collect a large number of fractions (say, fifty to one hundred 5 cm^3 fractions) in test tubes and would need to identify which tubes contained protein. However, we cannot see where proteins are just by looking! We need a method that will tell us where the protein(s) are (ie what tubes they are in) and approximately how much protein is present in each tube, so that we can produce an elution profile such as that shown in Figure 2.1.

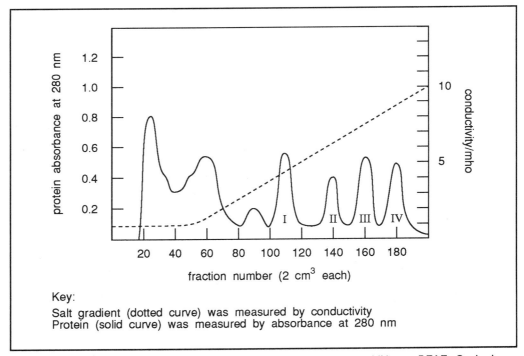

Figure 2.1 Chromatographic separation of rabbit muscle aldolase isozymes (I-IV) on a DEAE - Sephadex column (1cm x 25cm). The 2cm^3 fractions were eluted with a linear gradient from 0 to 0.35 mol l^{-1} NaCl in 0.01 mol l^{-1} Tris, 0.001 mol l^{-1} EDTA buffer solution (pH 7.5).

There are also other occasions where we need to know how much protein is in a given solution, for example, when determining the specific activity of a protein.

peptide bonds   Proteins and peptides are formed by linking amino acids via peptide bonds (Figure 2.2).

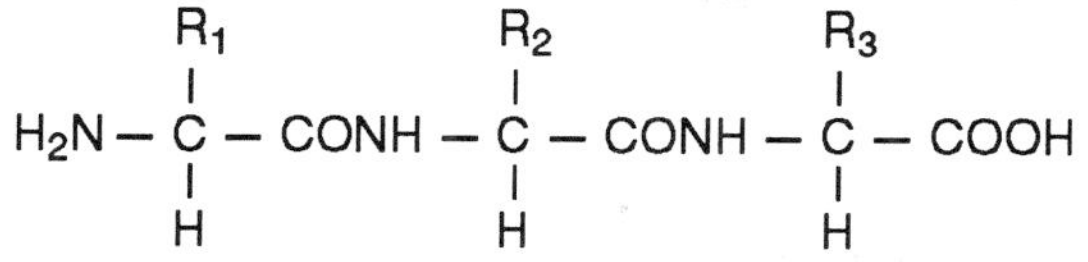

$$H_2N - \underset{\underset{H}{|}}{\overset{\overset{R_1}{|}}{C}} - CONH - \underset{\underset{H}{|}}{\overset{\overset{R_2}{|}}{C}} - CONH - \underset{\underset{H}{|}}{\overset{\overset{R_3}{|}}{C}} - COOH$$

Figure 2.2 A typical polypeptide chain (R_1, R_2, and R_3 are sidechains of amino acids).

To measure the presence of, and/or to quantify a mixture of polypeptides, one approach would be to devise a method that would measure the number of peptide bonds present which in turn will be a measure of the total amount of polypeptides present. Such a method would be both qualitative (ie it will tell us that there actually is protein present) and quantitative (it will tell us how much protein is present).

Other methods for detecting and measuring polypeptides rely on the ability of certain amino acid side chains ('R groups' in Figure 2.2) to give a coloured product on reaction with a particular reagent. Since the intensity of colour developed is dependent not on the total amount of protein, but on the amount of the particular amino acid present, most colorimetric tests tend to be qualitative rather than quantitative. The difference between qualitative and quantitative assays will be developed further in later sections.

We will now consider some of the more commonly used methods for measuring proteins in solution.

2.2 Direct spectrophotometric methods for measuring proteins in solution

non-destructive methods of protein analysis   Both the following methods are non-destructive. By this we mean that having made spectrophotometric measurements, the sample in the cuvette can be recovered and used further. This is particularly useful when one is working with small amounts of polypeptide and cannot afford to waste any.

2.2.1 Absorbance at 280 nm

The side chains of two of the twenty amino acids commonly found in proteins absorb UV radiation at 280 nm. These are tryptophan and tyrosine (their UV spectra are shown in Figure 2.3). The absorbance at 280 nm is therefore a simple method for detecting the presence of protein.

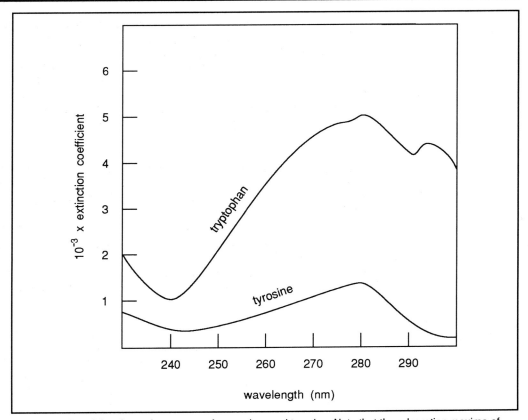

Figure 2.3 Ultraviolet absorption spectra of tryptophan and tyrosine. Note that the adsorption maxima of these two amino acids is close to 280 nm.

∏   However, it is not a particularly accurate method for determining protein concentration. Can you explain why?

The absorbance at 280 nm is dependent on the total amount of tyrosine and tryptophan present in the various proteins in a solution, rather than on the absolute amount of protein present. It is quite possible therefore to have two different protein solutions, each containing the same mass of protein but having different absorbance at 280 nm due to their differing contents of tyrosine and tryptophan.

However, in practice, the vast majority of pure proteins, have an extinction coefficient in the range 0.4 to 1.5 cm^2 mg^{-1} at 280 nm. Therefore for most protein mixtures it can be assumed to a good approximation that the solution has an extinction coefficient of about 1.0 cm^2 mg^{-1} at 280 nm.

How do we relate the absorbance at 280 nm to the concentration of protein? You may need to be reminded of the Beer-Lambert relationship:

$$A = \varepsilon c l \qquad \qquad \text{(E - 2.1)}$$

where A is the absorbance at a given wavelength, ε is the extinction coefficient (units = cm^2 mg^{-1}) at that wavelength, c is the concentration of the sample expressed as mg cm^{-3}

or mg ml^{-1} and l is the path length of the cuvette (normally 1 cm). Using this equation, we can deduce that the absorbance of a 1 mg cm^{-3} solution of protein at 280 nm is approximately 1.0 (since $\varepsilon = 1$ cm^2 mg^{-1} approx, $c = 1$ mg cm^{-3} and $l = 1$ cm). (NB: in the Beer Lambert Equation (E - 2.1) the molar absorption coefficient ε has the units m^2 mol^{-1} when all the terms are defined according to SI rules). You need to be a little careful with units to make sure that they are consistent with each other when you apply the Beer-Lambert relationship. In handling protein solutions, we often describe concentrations in terms of mg ml^{-1}. If you remember that mg ml^{-1} is equivalent to mg cm^{-3} you should not encounter any difficulty.

SAQ 2.1	You are provided with a solution of a crude protein extract. 0.1 cm^3 of this solution was added to 2.9 cm^3 of buffer solution and the absorbance of this solution at 280 nm was found to be 1.3 using a 1 cm path length cell. Calculate the approximate protein concentration of this solution.

Once the protein has been purified it is of course a simple task to determine an accurate extinction coefficient for the protein by preparing a solution of known mass concentration and measuring its absorbance. This extinction coefficient may then be used on later occasions to determine the concentration of this protein in other purified samples.

SAQ 2.2	An enzyme was purified to homogeneity. A 0.4 mg cm^{-3} solution of this enzyme was prepared and the absorbance at 280 nm recorded to be 0.25 using a 1 cm path length quartz cell. Calculate the extinction coefficient for this enzyme. In a later study 20 cm^3 of purified enzyme solution was prepared. This solution had an absorbance of 0.08 at 280 nm using a 1 cm path length cuvette. How much purified enzyme was present in the 20 cm^3 solution?

2.2.2 Absorbance at 220 nm

The peptide bond has an absorption maximum at 192 nm. A 1 mg cm^{-3} solution of proteins has a calculated absorbance of about 60 at 192 nm. Unfortunately, the spectrophotometers that we routinely use cannot be used at this low wavelength. However, the peak at 192 nm is a broad one and even at 220 nm (which is well within the working range of most spectrophotometers) a 1 mg cm^{-3} protein solution has a calculated absorbance of about 11.0 using a 1 cm path length cell. It should be noted that absorbance values in excess of 2 are meaningless and cannot be measured directly (most spectrophotometers have an absorbance scale of 0 - 2) - these high values have been obtained by calculation using absorbance values measured for solutions of much lower concentration.

∏ Can you remember what the absorbance of a 1 mg cm^{-1} solution of protein was at 280 mg? Compare this with the absorbance obtained by measuring at 220 nm. What does this tell you about the relative sensitivities of the two methods? In other words, is measuring at 220 nm more sensitive than measuring at 280 nm or *vice versa*?

The absorbance of a 1 mg cm^{-3} protein solution at 280 nm was about 1.0. Monitoring at 220 nm is therefore at least ten times more sensitive than monitoring at 280 nm. If we take an absorbance of 0.01 as being the lower limit of detection of most spectrophotometers, then measuring the absorbance at 220 nm is capable of detecting proteins down to a concentration of about 0.001 mg cm^{-3} (1 μg cm^{-3}).

There are two further advantages of measuring at 220 nm:

- peptides and proteins that contain no tyrosine or tryptophan will be detected at 220 nm, whereas they will not be detected at 280 nm. This is particularly important when monitoring peptides, since it is quite likely that many will be devoid of tyrosine and tryptophan;

- by measuring at 220 nm we are measuring the peptide bonds (this is generally true, although there is a small contribution to the absorbance from some amino acid side chains). This method is therefore a reasonably quantitative method for determining the amount of protein. If we have two solutions, one (solution A) having twice the concentration of protein of the other (solution B), we will have double the amount of peptide bonds in sample A over sample B and therefore the absorbance for sample A will be approximately twice that of sample B.

Measuring the absorbance at 220 nm therefore seems far superior in every way to measuring absorbance at 280 nm. Why, therefore, did we even bother to mention measuring the absorbance at 280 nm? Unfortunately, many of the buffer solutions and reagents, such as detergents, that we use in protein purification methods themselves absorb at 220 nm (but not at 280 nm). Therefore, it is not always possible to measure at 220 nm and often you will find workers making measurements at 280 nm.

Note: it is essential to use quartz cuvettes when making measurements at both 220 and 280 nm. Quartz is transparent to UV radiation whereas glass and plastic cuvettes are not.

SAQ 2.3

Figure 2.4 a) shows an ion-exchange column elution profile monitored by absorbance at 280 nm. Two proteins, A and B, have been eluted and appear to contain approximately equal amounts of protein since the peaks are of approximately equal size.

Figure 2.4 b) shows the same elution profile monitored by absorbance at 220 nm. By comparing peak size, there now appears to be about five times as much protein in peak A as there is in peak B.

Can you explain the apparently different relative amounts of A and B in the two profiles? In terms of measuring the relative amounts of protein in A and B, which is likely to be the more accurate? Figure 2.4 b) also shows an extra peak, C, which was not present in Figure 2.4 a). Amino acid analysis showed peak C to be a polypeptide. Can you explain why this peak should appear in one figure and not the other?

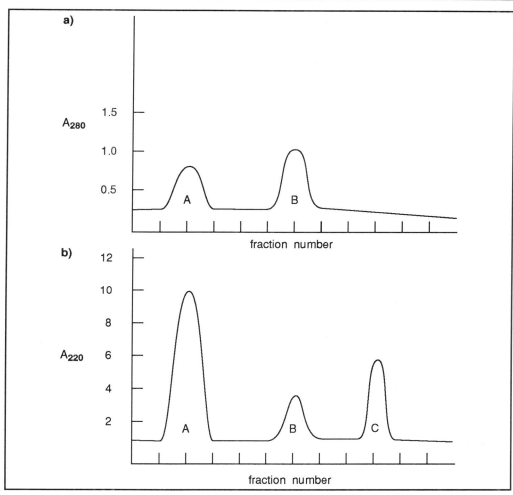

Figure 2.4 Idealised elution profiles of a mixed protein solution monitored at a) 280 nm and b) 220 nm. (NB The absorbance at 220 nm were measured on diluted samples and values were extrapolated to the same concentrations used for a). Note that in practice, elution peaks are usually not quite so symmetrical. Unevenness in the packing of the ion exchange column and thermal diffusions frequently produces distortions in the elution profiles.).

2.3 Colorimetric methods

destructive methods of protein analysis

Three colorimetric methods have found widespread use by protein chemists and will be described here. They are all 'destructive' methods in that the protein being measured is not recoverable.

2.3.1 The Biuret method

The Biuret reagent is an alkaline copper sulphate solution containing sodium potassium tartrate. When mixed with a protein solution the copper (II) ions form a co-ordination complex with four-NH peptide bond groups (Figure 2.5) giving a blue colour with an absorbance maximum at 540 nm. One simply therefore mixes the protein solution with the reagent and measures the absorbance at 540 nm, the increase in absorbance over an appropriate blank is a measure of the amount of protein present.

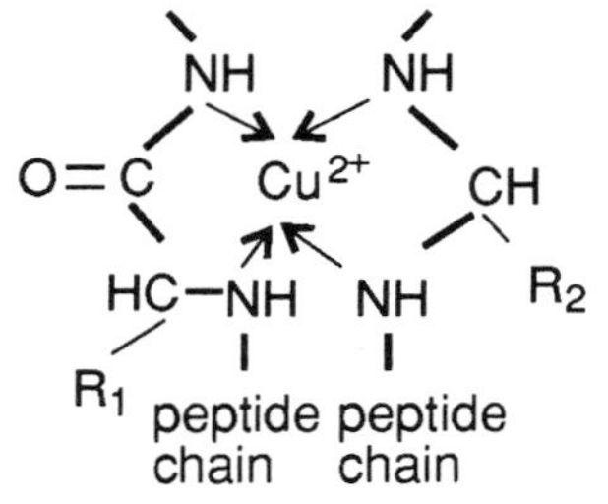

Figure 2.5 Co-ordination complex formed between copper (II) ions and peptide bonds in the Biuret assay.

Π   Would you consider this to be a very quantitative method for measuring proteins?

Since it is effectively measuring the presence and number of peptide bonds, it is a reasonably quantitative assay. Unfortunately it is not a very sensitive method, needing protein concentrations of at least 1 mg cm^{-3}. Compare the sensitivity of this method with the methods of measuring absorbance at 220 and 280 nm, and the other colorimetric methods described below.

2.3.2 The Lowry (Folin Ciocalteau) method

This method combines the copper reaction of the Biuret method with the Folin-Ciocalteau reagent which reacts with tyrosine residues in proteins. On mixing the Lowry reagent with a protein solution a dark blue/purple colour, with maximum absorbance at 660 nm, is produced.

Since much of the colour developed is due to the presence of tyrosine residues, care must be taken when using this method to produce quantitative data. For example, two pure proteins, each at 1 mg cm^{-3} but differing in their tyrosine content, will give different colour intensities. However, when the method is applied to relatively complex protein mixtures, the tyrosine content of different mixtures tends to approximate to a similar average figure, so the method can be used to compare the protein content of mixtures fairly accurately. The method is quite sensitive, being sensitive down to 50 μg cm^{-3}, and has found wide usage.

2.3.3 The Bradford method

This method uses the dye, Coomassie Blue G250. An acid solution of the dye is red-brown and has an absorption maximum at 470 nm. When mixed with a protein solution, the dye binds to the protein (predominantly to the arginine residues, but also partly to aromatic residues) and the absorption maximum shifts to 595 nm. The increase in absorbance at 595 nm, relative to an appropriate blank, is therefore a measure of the amount of protein present.

However, since the intensity of colour developed is proportional to the amount of arginine (and to a lesser extent to the amount of aromatic residues) in a protein sample, care must again be taken in using this assay for providing quantitative data. Two purified proteins, of equal concentration, need not necessarily give the same colour intensity with the Bradford reagent. However, when analysing complex mixtures of proteins, the arginine content of different mixtures tends to approximate to a similar average figure, so the method is frequently used to compare the protein content of protein mixtures. The method is sensitive down to concentrations of 1-10 μg cm^{-3} and is the method of choice nowadays. Coomassie Blue is also used to detect and measure proteins in electrophoretic gels. In this case the method is usually called Coomassie staining. We will examine this in Section 2.5.

SAQ 2.4

A series of dilutions of bovine serum albumin (BSA) were prepared and 0.1 cm^{-3} of each solution subjected to a Bradford assay. The increase in absorbance at 595 nm relative to an appropriate blank was determined in each case, and the results are shown in the table below:

Concentration of BSA (mg cm^{-3})	A_{595}
1.5	1.40
1.0	0.97
0.8	0.79
0.6	0.59
0.4	0.37
0.2	0.17

Plot a calibration graph of BSA concentration against A_{595}. 0.1cm^{-3} of a solution of unknown BSA concentration gave an A_{595} of 0.98 in the same assay. Using the graph, determine the concentration of BSA in this unknown sample. At the same time as doing this work another worker in the laboratory had just purified a new protein 'B'. On the same day of the calibration, he committed 0.1 cm^{-3} of his purified protein solution to a Bradford assay using the same reagents and protocol and obtained an A_{595} of 0.5. What can you deduce about the concentration of this new protein 'B'?

2.4 Amino acid analysis

amino acid analysis

Probably the most accurate way to determine the total amount of protein in a solution is to carry out an amino acid analysis on part of the sample. However, while extremely accurate, it is time-consuming, taking normally two working days to get a result, compared with a few minutes for the spectrophotometric methods and 30 minutes to 1 hour for the colorimetric methods.

To carry out an amino acid analysis, the sample to be analysed is firstly dried in a glass tube. Subsequently, 6 mol l^{-1} HCl is added, the tube is sealed and then heated at 105° for 24 hours. This treatment hydrolyses all the peptide bonds in the polypeptides and converts all polypeptides to a mixture of their constituent amino acids. Following this hydrolysis step, the sample is dried under vacuum to remove the HCl, dissolved in a low pH buffer solution (to ensure all the amino acids are positively charged) and loaded onto an ion exchange column, typically a negatively-charged sulphonated polystyrene resin (see Section 1.3.4). All the amino acids bind to the column and are then eluted sequentially by applying a shallow salt and/or pH gradient. As the various amino acids emerge from the bottom of the column they need to be identified and quantified. This is achieved by mixing ninhydrin solution with the column effluent and passing this effluent through a heating coil at 95°C. At this stage the ninhydrin reacts with each amino acid to give a blue colour (yellow for imino acids) which is then detected and quantified by passing the effluent stream through a spectrophotometer flow cell. A chart printout records the individual peaks as they are measured. The identity of each amino acid is established on the basis of the position in the chromatogram. The process is shown diagrammatically in Figure 2.6.

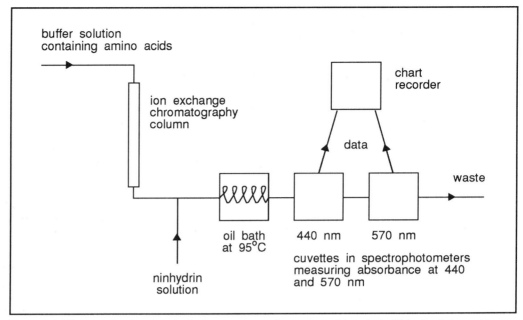

Figure 2.6 Diagram showing the principal components of an amino acid analysis. Note: the imino acids, proline and hydroxyproline, give a yellow colour with ninhydrin and are read on 440 nm; the remainder of the amino acids give a blue colour with ninhydrin and are read at 570 nm.

The system is initially calibrated by running a standard mixture of amino aids of known concentration. The peak area produced by a known amount of each amino acid is determined and this value is then used to convert the peak areas for each amino acid in the test sample to the mass of that amino acid. The total sum of the mass of each amino acid in the sample gives the total mass of protein originally present.

We have described above the use of ninhydrin to detect the amino acids. This method has been used successfully for many years, but recent attempts to increase sensitivity have resulted in ninhydrin being replaced by reagents such as o-phthalaldehyde, which also reacts with the amino groups of amino acids, but gives a fluorescent, rather than coloured, derivative that is measured by a fluorometer. Methods such as this have greatly improved the sensitivity of amino acid analysis, with some workers claiming sensitivity down to as low as 100 femtomole (1 fmol = 10^{-15} mol).

∏   Assume you have purified a protein of molar mass 50 000 g, what mass of protein would you need to provide 100 fmoles of your protein for amino acid analysis?

The mass of 1 mol = 50 000 g

∴ 1 fmol has a mass of $50\,000 \times 10^{-15}$ g

$= 50 \times 10^{-12}$ g $= 50$ pg (50 picogrammes).

Thus you would need 5000 pg of protein.

This really is not a lot of protein! The development of highly sensitive methods of analysis has been a feature of protein and peptide analysis in recent years. You will come across the analysis of exceedingly small amounts of protein again when you read about protein sequence determination.

2.5 Gel electrophoretic methods

polyacrylamide gel

The methods described previously tell you how much protein is present in a solution; it does not tell you how many different proteins are present. If your Biuret test tells you that you have 1 mg of protein in a 1 cm^3 (l ml) of solution, this does not tell you if you have 1 mg of pure protein or 1 mg of total protein comprising hundreds of different proteins. To answer this question we need to separate out and visualise the individual proteins. Traditionally this has been done by electrophoresis (Section 1.3.4). Over the years a number of different support media have been used for electrophoresis (eg paper, starch, nitrocellulose) and although these still have their uses, the most commonly used support nowadays is the polyacrylamide gel. Polyacrylamide gel electrophoresis (often abbreviated to 'PAGE') is the workhorse of the protein chemist and is used almost daily to monitor protein purification.

Polyacryamide gels are prepared by cross-linking the small acrylamide monomer molecule in the presence of a small amount of the cross-linking molecule N,N'-methylenebisacrylamide (see Figure 2.7).

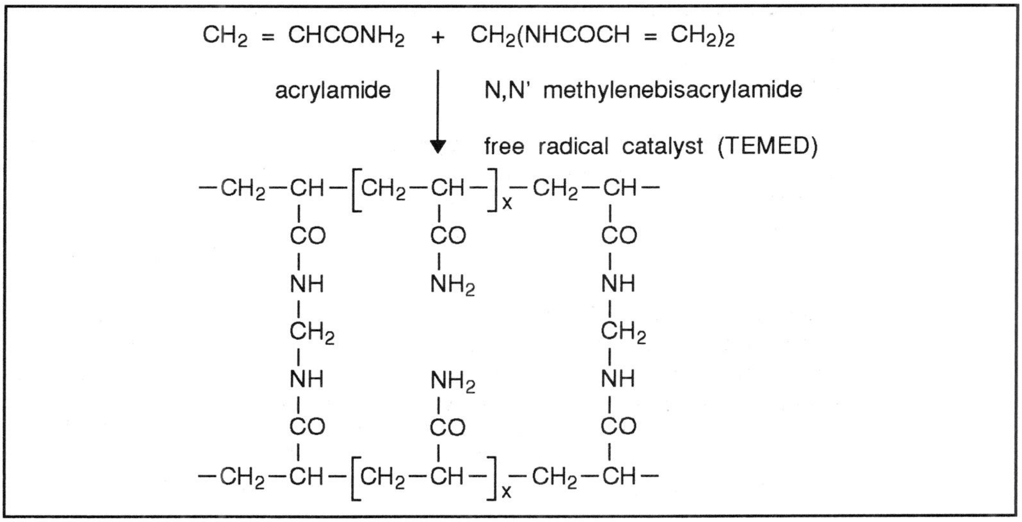

$$CH_2 = CHCONH_2 \quad + \quad CH_2(NHCOCH = CH_2)_2$$

acrylamide   N,N' methylenebisacrylamide

free radical catalyst (TEMED)

$$-CH_2-CH-\left[CH_2-CH-\right]_x- CH_2-CH-$$

Figure 2.7 Polymerisation of acrylamide.

The cross-linking is normally done by chemical polymerisation using an initiator such as ammonium persulphate and the catalyst N,N,N'N'-tetramethylenediamine (TEMED). The final polymerised gel should be considered as a cross-linked three-dimensional matrix containing 'pores' through which the proteins will pass. The size of these 'pores' is obviously a function of the amount of cross-linking (the greater the amount of cross-linking the smaller the pore size and *vice versa*, this can be controlled by varying the amount of acrylamide and bis-acrylamide in the polymerisation solution. The reasons for wanting to do this will become apparent later. Acrylamide gels are usually prepared as 'slab' gels between two glass plates. A plastic 'comb' is placed in the top of the gel solution. When the gel has set this comb is removed to reveal a series of loading wells adjacent to one another. In this way, up to twenty samples can be loaded on a single gel (see Figure 2.8).

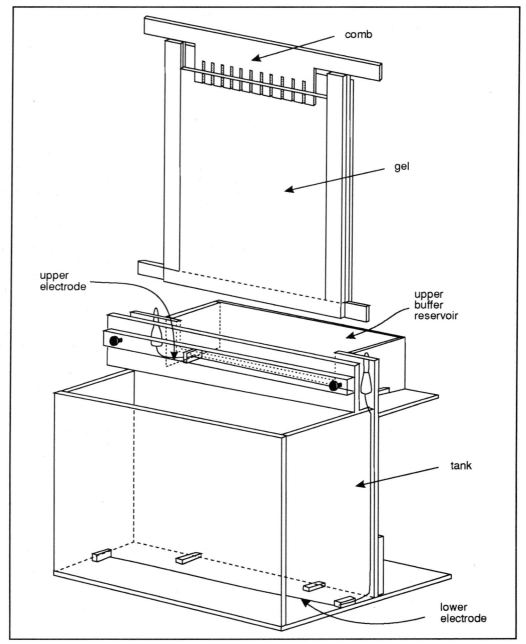

Figure 2.8 A typical vertical gel apparatus. Internal gel plate size: 13 cm length x 11cm wide. Spacer thickness: 1.5-2.0 mm. In operation, the gel is inserted into the tank so that its lower part is immersed in the layer of buffer which covers the lower electrode. The upper surface of the gel protrudes into the upper buffer reservoir. Buffer in this compartment covers the upper electrode and fills the 'wells' in the gel produced by the removal of the comb. The protein mixtures are layered carefully into the wells. Then an electropotential is applied across the gel by attaching the electrode terminals to a 'power pack' which controls the voltage difference between the two electrodes. Further details are given in the text.

The most commonly used gel system for studying proteins is SDS gel electrophoresis, a typical example of which is shown in Figure 2.9

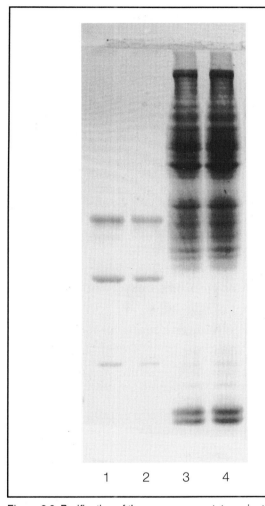

Tracks 3 and 4, total *E. coli* protein extract. Tracks 1 and 2, the result of ammonium sulphate precipitation and two ion-exchange chromatography steps on the crude extract. The sample now contains three proteins; the middle of the three bands corresponds to the enzyme aspartate aminotransferase. Note that the enzyme is present at a relatively low concentration in the crude extract; at the loadings used the enzyme cannot be seen in the tracks of the crude extract.

Figure 2.9 Purification of the enzyme aspartate aminotransferase from *E. coli*, SDS gel electrophoresis.

The protein-containing sample to be analysed is first heated in the presence of the anionic detergent sodium dodecyl sulphate (SDS): $CH_3(CH_2)_{10}CH_2OSO_3^- Na^+$ and 2-mercaptoethanol. The aim of this step is to completely denature (unfold) the protein chain.

∏ Why do you think the mercaptoethanol is present?

One of the forces that holds the three-dimensional structure of some proteins together is the disulphide bridge, a covalent linkage between two cysteine residues. The 2-mercaptoethanol reduces any disulphide bridges present in the protein (the reduction of disulphide bridges is shown in Figure 2.10).

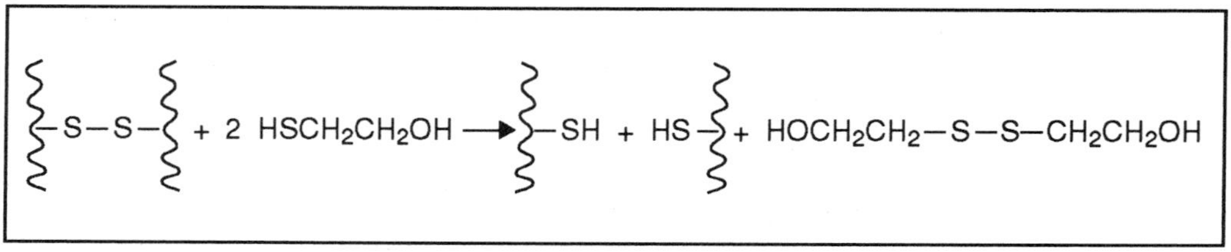

Figure 2.10 Disulphide bridge reduction with 2-mercaptoethanol.

The SDS binds to the polypeptide chains through hydrophobic interactions, resulting in the binding, on average, of one negatively charged SDS molecule to every two amino acids. The consequence of this treatment is that every polypeptide in the mixture is completely denatured and opened out into a rod-shaped structure with a series of negatively charged SDS molecules along the polypeptide. The original native charge on the molecule is completely swamped by the SDS molecules. The mixture is then loaded into one of the wells of the gel (the wells of the gel are already filled with an appropriate buffer solution) together with an ionisable tracking dye, usually bromophenol blue, that allows the electrophoretic run to be monitored. When a potential gradient is applied, since all the polypeptides now have a net negative charge, they will move towards the positive electrode. As the polypeptides move through the gel they separate.

∏ Can you work out the principle behind this separation?

You would be wrong if you assumed that a larger protein travels faster because it has a greater number of negative charges on it than has a smaller protein. It does indeed have a greater number of charges on it, but the charge per unit length, on each protein is the same, so they are driven with an equal force per unit length into the gel. However, the smaller protein will travel more quickly through the gel than the large protein.

They separate according to size due to the sieving effect of the gel pores. Quite simply, the smaller the molecule the easier it can pass through the pores of the gel, whereas the larger the molecule the more resistance it will experience and the longer it will take to travel through the gel. Separation of polypeptides on an SDS gel is therefore based on size.

When the tracking dye has reached the bottom of the gel, electrophoresis is stopped, the gel is removed from between the glass plates and immersed in an appropriate protein stain solution (usually Coomassie brilliant blue) for a few hours; it is then washed in a destain solution overnight.

Although initial observations can be made the same day as the gel is run, a complete analysis of the gel is usually made the following day. An initial crude extract will of course show a complex pattern of many bands (see Figure 2.9) but as one purifies the protein, so the gel pattern should simplify until eventually a single band is observed on the gel, indicating that the protein is now pure (Figure 2.9).

∏ Sometimes a pure protein may give more than one band on a SDS gel. Can you think why this may happen?

This will occur when a protein consists of two or more dissimilar subunits. The SDS/mercaptoethanol treatment disrupts the protein into its subunits and each individual subunit will be seen on the SDS gel.

The gel shown in Figure 2.9 is a 15% polyacrylamide gel; ie total acrylamide and bis-acrylamide concentration is 15 g per 100 cm^3 (ml). This separates proteins in the relative molecular mass range 10 000 (these proteins run near the dye front at the bottom of the gel) to approximately 100 000 (these proteins just enter the gel). Many proteins that we study come in this range and therefore 15% gels are frequently used by many workers. However, supposing you were studying a protein of relative molecular mass (RMM) 160 000; where would this protein run on a 15% polyacrylamide gel? The answer is that it would not run. The pores of this gel are too small for it to enter the gel and it would remain at the top of the gel. This is most unsatisfactory since you cannot tell whether a band at the top of your gel comprises only one or a dozen different proteins. In this case you would need to use a gel with a larger pore size (ie lower cross-linking achieved by using a lower acrylamide concentration) in order to resolve and visualise these proteins. For this reason some workers necessarily use 10% or even 7.5% gels for studying their larger proteins.

The technique of SDS-gel electrophoresis can also be used to obtain an approximate value for the RMM of a protein (see Chapter 1).

| SAQ 2.5 | We have come across the dye Coomassie brilliant blue twice in this section. Can you remember what is was used for? |

| SAQ 2.6 | List as many methods as you can for detecting the presence of protein in a solution. |

| SAQ 2.7 | Name the two amino acids that are responsible for UV absorption by proteins at 280 nm. |

SAQ 2.8

State whether each of the following statements is true or false:

1) in polyacrylamide gels, the higher the concentration of acrylamide, the smaller the pore size in the gel;

2) SDS gel electrophoresis separates proteins according to their charge;

3) a protein cannot be pure if it shows two bands on a SDS polyacrylamide gel;

4) the Bradford method is used to stain proteins on acrylamide gels;

5) there are no methods that will detect protein at concentrations of less than 5 $\mu g\ cm^{-3}$;

6) in amino acid analysis, amino acids are separated according to their size;

7) a 1 mg cm^{-3} solution of protein has an absorbance of about 10 at 280 nm, when measured in a 1 cm cell.

| SAQ 2.9 | Name one non-destructive and two destructive methods for estimating proteins in solution. |

Summary and objectives

Various physical and chemical methods to detect and estimate proteins are described and their advantages and disadvantages discussed. The most rapid routine method is the non-destructive spectrophotometric measurement of the absorbance at 220 nm (due to the presence of peptide bonds) and/or 280 nm (due to the presence of tryptophan or tyrosine residues).

The destructive methods involve the formation of coloured complexes with the copper II ion (Biuret and Lowry methods) or Coomassie blue (Bradford method).

Amino acid analysis, using ion exchange columns, provides a quantitative estimation of individual amino acids present in the protein; it also gives information about the total amount of protein present. This method is not suitable for rapid routine studies.

Polyacrylamide gel electrophoresis (PAGE) studies are used for the separation of proteins in mixture. SDS gel electrophoresis enables estimates of relative molecular mass to be made.

On completion of this chapter you should be able to:

- classify the various methods for routine protein assay and explain the underlying principles of each;

- discuss the relative advantages of measuring the absorbance of a protein solution at 220 and 280 nm;

- describe the hydrolysis of a protein and the quantitative analysis of the amino acids present in the hydrolysate;

- explain the principles of polyacrylamide gel electrophoresis (PAGE) for the separation of proteins in a mixture.

Protein purification

Protein purification

3.1 Introduction

At first sight, the purification of a protein from a crude extract such as a tissue homogenate seems a daunting task. One has to isolate a single protein from a mixture of possibly five to ten thousand other proteins. (A recent survey of 100 published purification protocols showed that on average the protein of interest was present at 0.016% in the initial extract, ie 1 part in 7000). How many separate purification steps do you think it takes to purify a single protein from such a crude homogenate? You may be surprised to know that the majority of published protocols for the purification of proteins involve about four steps, and some proteins have been purified using as little as two steps. Protein purification is not therefore the awesome task that it may seem at first glance. The basic strategy in any protein purification is to exploit the different physical and chemical properties of the proteins in a mixture.

∏ Make a list of the ways proteins in a mixture may differ from one another so that we can exploit these differences in developing purification methods. To start you off, there will obviously be a range of proteins with different molecular masses, so the separation of proteins according to size is one possible approach and, indeed is frequently used.

Now refer to the various physico-chemical parameters, listed below, that are used in protein purification. Do not worry if you did not get them all as some of these terms are likely to be unfamiliar to you. They will all be described in some detail later.

The physico-chemical differences which can be used to separate proteins are:

- stability;

- solubility;

- size;

- charge;

- hydrophobicity;

- ligand binding.

However, to fully understand how we can exploit their differences it is necessary to first ensure that you understand some general principles of protein structure.

3.2 Protein structure

3.2.1 Isoelectric point and the variation of the charge on proteins with pH

α-amino acids   Proteins consist of chains of α-amino acids linked by peptide bonds (see Section 3.2.2). The general structure of an α-amino acid is often represented as in Figure 3.1.

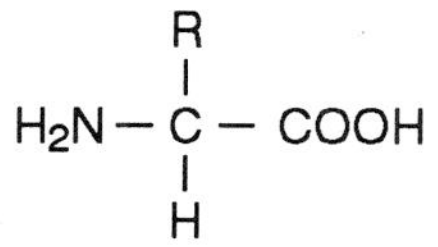

$$R$$
$$|$$
$$H_2N - C - COOH$$
$$|$$
$$H$$

Figure 3.1 General formula of α-amino acids.

∏   Explain why the term α-amino acid is used to describe this structure.

This molecule contains an amino (-NH$_2$) group and an acid carboxyl (-COOH) group, hence the name amino acid. They are α-amino acids because the amino group is linked to the carbon atom adjacent to the COOH group and which, by convention, is designated α. The R group varies depending on which amino acid we are describing.

Proteins are made up of a mixture of twenty possible amino acids (ie there are 20 different R groups) and these are shown in Table 3.1. This is a complex table and contains a lot of information. We anticipate that you will be familiar with many of the amino acids from your previous studies. We do not expect you to remember the exact RMMs values for each of the amino acids (you can work these out if you know the structures of the amino acids) nor their pK values.

If you examine Table 3.1 carefully you will see that each amino acid has a low pK value (at about pH 2) which represents the ionisation of the α-carboxylic acid group and another pK at about pH 9-9.5, which corresponds to the ionisation of the amino group. Amino acids with additional ionisable groups (for example aspartic acid) have additional pK values.

To encourage you to read Table 3.1 carefully, put a tick beside the amino acids with ionisable R groups and a cross beside those amino acids with hydrophobic side chains. You will be able to check your response later.

Note that the amino acids are often referred to by an abbreviated form. For example glycine is often written as Gly. Increasingly, particularly as more and more amino acid sequences of proteins have become established, a single letter notation has been used to identify individual amino acids. For example glycine has been given the letter G. This obviously simplifies writing down long amino acid sequences. We have provided a list of the various abbreviations used for the common amino acids in an appendix at the end of this text. The three-letter abbreviations are usually fairly obvious and easy to remember. The single-letter symbols are less easy to remember.

Name (abbreviation)	R	RMM	pK values (pI)	Occurrence[a]	General comments[b]
Monoaminomonocarboxylic acids					
glycine (gly)	H-	75.05	2.34, 9.60 (5.97)	M: silk, gelatin	only amino acid without asymmetric carbon
L-(+)-alanine (ala)	CH_3-	89.10	2.35, 9.69 (6.02)	M: silk, gelatin, zein	N
L-(+)-valine (val)	$(CH_3)_2CH$-	117.15	2.32, 9.62 (5.97)	S: range of fibrous proteins	E, N
L-(-)-leucine (leu)	$(CH_3)_2CHCH_2$-	131.18	2.36, 9.60 (5.98)	M: haemoglobin, zein	E, N, one of the most common amino acids
L-(+)-isoleucine (ile)	$CH_3CH_2CH(CH_3)$-	131.18	2.36, 9.68 (6.02)	S: casein, zein	E, N
L-(-)-serine (ser)	$HOCH_2$-	105.10	2.21, 9.15 (5.68)	S: silk and range of proteins	H
L-(-)-threonine (thr)	$CH_3CH(OH)$-	119.12	2.09, 9.10 (5.60)	S: casein	E, H
Monoaminodicarboxylic acids					
L-(+)-aspartic acid (asp)	$HOOCCH_2$-	133.11	2.09, 3.86, 9.82 (2.98)	S: range of proteins	D, widely distributed in plants
L-(+)-glutamic acid (glu)	$HOOC(CH_2)_2$-	147.13	2.19, 4.25, 9.67 (3.22)	M: prolamines, globulins, casein	D
L-(-)-asparagine (asn)	H_2NOCCH_2-	132.13	2.02, 8.80 (5.41)		
L-(+)-glutamine (gln)	$H_2NOC(CH_2)_2$-	146.15	2.17, 9.13 (5.70)	range of plants, roots, seedlings	
Diaminomonocarboxylic acids					
L-(+)-lysine (lys)	$H_2N(CH_2)_4$-	146.19	2.18, 8.95, 10.53 (9.74)	M: albumins, globulins	E
L-(+)-arginine (arg)	$H_2NC(NH)NH(CH_2)_3$-	174.20	2.17, 9.04, 12.48 (10.76)	M: protamines, histones	E, hydrolysis gives ornithine
diaminopimelic acid (dap)	$HOOCCH(NH_2)(CH_2)_3$-	190.05			only found in bacterial proteins
L-(+)-ornithine	$H_2N(CH_2)_3$-	132.16	1.94, 8.65 (9.7)		isolated from proteins after alkaline hydrolysis

Table 3.1 Properties of naturally occurring amino acids ($RCH(NH_2)COOH$).
Key: [a]M, major component; S, small amounts. [b]E, essential amino acid; N, non-polar sidechain acts as spacer in proteins; H, -OH group forms bonds with -NH and -CO group of main protein chain; D, -$CONH_2$ acts as a donor acceptor of hydrogen bonds.

Name (abbreviation)	R	RMM	pK values (pI)	Occurrence[a]	General comments[b]
Sulphur-containing amino acids					
L-(+)-cysteine (cys)	$HSCH_2-$	121.16	1.71, 8.33 (5.02) 10.78	S: range of proteins	breakdown product of cystine in body: soln oxidised to cystine
L-(-)-cystine (cys-cys)	SCH_2- SCH_2-	240.30	1.0, 2.1, 8.02, 8.71 (5.06)	abundant in skeletal tissue, hair	-S-S-bonds play important role in protein structure stabilisation
L-(-)-methionine (met)	$CH_3S(CH_2)_2-$	149.21	2.28, 9.21 (5.75)	egg and vegetable proteins	E, important role in biological methylations
Aromatic amino acids					
L-(-)-phenylalanine (phe)	$C_6H_5CH_2-$	165.19	1.83, 9.13 (5.48)	ovalbumin, zein, fibrin	E
L-(-)-tyrosine (tyr)	$p-HOC_6H_4CH_2-$	181.19	2.20, 9.11, 10.07 (5.67)	wide range of proteins	H, least soluble of all amino acids
Heterocyclic amino acids					
L-(-)-tryptophan (trp)	CH_2- (indole)	204.22	2.38, 9.39 (5.88)	globulins, albumins, casein	E,H, not synthesised by human body
L-(-)-histidine (his)	CH_2- (imidazolium)	155.16	1.78, 5.97, 8.97 (7.59)	histones, animal and vegetable globulins	E (rats)
Imino acids					
L-(-)-proline (pro)	(pyrrolidine structure)	115.13	1.99, 10.6	wide range of proteins	
L-(-)-hydroxyproline (hyp)	(hydroxypyrrolidine structure)	131.13	1.82, 9.65	collagen, gelatin	not strictly an amino acid; extracted from protein hydrolyzates

Table 3.1 (continued) Properties of naturally occurring amino acids (RCH(NH₂)COOH).
Key: [a]M, major component; S, small amounts. [b]E, essential amino acid; N, non-polar sidechain acts as spacer in proteins; H, -OH group forms bonds with -NH and -CO group of main protein chain; D, -CONH₂ acts as a donor acceptor of hydrogen bonds.

The structure shown in Figure 3.1 is not a correct representation of how amino acids exist in solution. Figure 3.2 is a more realistic representation indicating as it does that amino acids are charged at all pH values. (We will deal with the ionisation of the R-groups later).

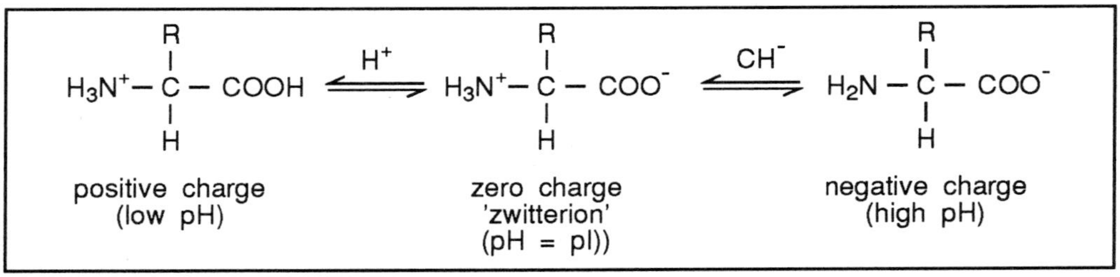

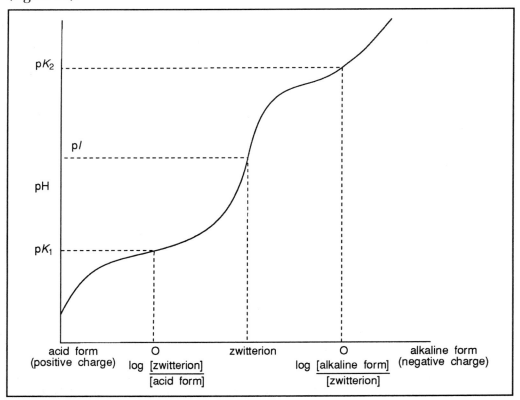

$$H_3N^+ - \underset{\underset{H}{|}}{\overset{\overset{R}{|}}{C}} - COOH \quad \underset{}{\overset{H^+}{\rightleftharpoons}} \quad H_3N^+ - \underset{\underset{H}{|}}{\overset{\overset{R}{|}}{C}} - COO^- \quad \underset{}{\overset{CH^-}{\rightleftharpoons}} \quad H_2N - \underset{\underset{H}{|}}{\overset{\overset{R}{|}}{C}} - COO^-$$

positive charge (low pH)   zero charge 'zwitterion' (pH = pI))   negative charge (high pH)

Figure 3.2   Ionisation of the α-amino acid and α-carboxylic acid groups of amino acids..

It is this charged nature that gives amino acids their characteristic properties, eg water solubility, crystalline structures etc. Each amino acid has an isoelectric point (pI), ie the pH value at which the amino acid has an overall zero charge. At this pI it does, however, have charged groups on it, it is just that the overall positive and negative charges balance each other out. This species is often referred to as a 'zwitterion'.

zwitterion

As the pH falls below this pI value (Figure 3.3), by adding H$^+$ ions the –COO$^-$ group, and the amino group become increasingly protonated to give an -$^+$NH$_3$ group (Figure 3.2).

Figure 3.3 Typical titration curve for an amino acid.

Ⲡ What therefore happens to the charge on an amino acid when we progressively reduce the pH of the solution from its isoelectric point?

Obviously the molecule becomes increasingly positively charged.

Do not be confused into thinking that the change from zero to positive charge is an all or nothing sudden one. There will always be an equilibrium between uncharged and charged species but the further the pH moves away (decreases) from the pI the greater the proportion of molecules that will be protonated and therefore have a net positive charge. This is illustrated in the titration curve (Figure 3.3).

Similarly, if we increase the pH from the isoelectric point by the addition of OH$^-$ ions we get an overall negative charge on the amino acid. In addition a number of amino acids also have charged R groups which also obviously contribute to the overall charge on the molecule, but this does not affect the principle that each amino acid has an isoelectric point, pI, (Table 3.1).

SAQ 3.1

From the information in Table 3.1 list the amino acids which will be negatively charged at pH 8.5.

Equally, therefore, since proteins are simply chains of linked amino acids, the same principle applies to proteins, namely that each individual protein has an isoelectric point (pI) at which it has an overall zero charge (Table 3.2). As the pH is reduced below this value it has an increasingly positive charge and conversely, as the pH is increased above this value, the protein has an increasingly negative charge. It is important that you understand this basic idea of the variation of charge on protein with pH on either side of its pI value, as it is central for the understanding of a number of purification procedures.

Protein		pI*
pepsin	approximately	1.0
fetuin		3.4
α-casein		4.0
ovalbumin		4.6
gelatin		4.8
serum albumin (human)		4.8
unrease		5.1
β-lactoglobulin		5.2
insulin		5.35
fibrinogen		5.5
catalase		5.6
carboxypepidase		6.0
collagen		6.6
myoglobin		6.99
haemoglobin (horse)		6.9
haemoglobin (human)		7.1
haemoglobin (chicken)		7.2
ribonuclease		9.5
cytochrome c		10.65
lysozyme		11.1

Table 3.2 Isoelectric points of some proteins. *The precise value depends on the temperature and ionic strength of the solution.

3.2.2 The function and location of amino acid side chains

peptide bonds

Proteins are formed by linking amino acids in chains through the formation of secondary amide linkages between the α-carboxyl and the α-amino groups of adjacent amino acids. Such bonds (CONH) are commonly known as peptide bonds (Figure 3.4).

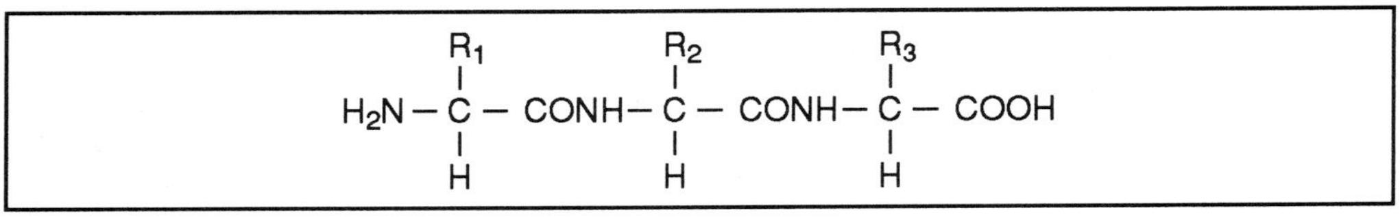

$$H_2N - \underset{\underset{H}{|}}{\overset{\overset{R_1}{|}}{C}} - CONH - \underset{\underset{H}{|}}{\overset{\overset{R_2}{|}}{C}} - CONH - \underset{\underset{H}{|}}{\overset{\overset{R_3}{|}}{C}} - COOH$$

Figure 3.4 The formation of peptide bonds between amino acids.

This chain is then folded into an overall three-dimensional shape. For many proteins, especially enzymes, the shape is generally globular. Proteins involved in structural roles, tend to be fibrous in shape. Picture, therefore a folded backbone chain but with different R groups pointing out along the backbone. Some of these R groups will be buried on the inside of the globular protein, others will be pointing outwards on the surface of the protein. For our purposes, many of the R groups fall into two important categories, they are either hydrophilic (charged or polar) or hydrophobic.

∏   Look at Table 3.1 and list the groups in the side chains (R groups) that can be described as charged, and those that can be described as hydrophobic (literally water hating).

The groups in the side chain which can be classified hydrophilic (water loving) groups are carboxyl and amino; the hydrophobic groups include those with aliphatic or aromatic character without a charged group.

We will deal with each of these types in detail.

hydrophilic groups

The hydrophilic groups are the groups in the side chains (R) that can be potentially charged in proteins depending on the pH of the solution. These include those from aspartic and glutamic acid, which have negatively-charged carboxyl groups on their side chains, and those from the three basic amino acids, lysine, arginine and histidine which are all capable of being protonated to give a positively-charged side chain.

Note that the side chain of lysine ends in an amino group. To differentiate this from the α-amino group of this amino acid it is referred to as the ε (epsilon) amino group or ε-NH$_2$. Can you see why it is given this name?

An easy one for Greek scholars! Continuing to letter the carbon atoms in the side chain from the α-carbon (attached to -COOH) atom in order β,γ,δ,ε, we see that the amino group is attached to the ε-carbon atom and is therefore the ε-amino group. Similarly, the side chain -COOH group in glutamic acid is referred to as the γ-carboxyl. Most chemical and biochemical nomenclature is based on such simple principles and you should not be daunted by nomenclatures that at first glance may seem quite alien.

hydrophobic groups

Typically hydrophobic groups are those from the amino acids alanine, valine, leucine and isoleucine (which have aliphatic side chains) and phenylalanine and tyrosine which have aromatic side chains). The heterocyclic amino acid, tryptophan is also hydrophobic in nature. Because of the presence of these 'water hating' side chains, not surprisingly these amino acids tend to be the least water soluble of all the amino acids.

∏ Amino acids, peptides and proteins are organic molecules. Many organic molecules, eg benzene and chloroform can be separated from mixtures by their volatility, ie by fractional distillation. Why is volatility not a potential method for separating proteins?

Although a number of the amino acid side chains are comprised of potentially volatile structures, eg aromatic and aliphatic hydrocarbon chains, the overwhelming characteristic of proteins is their ionic, charged nature. Such compounds are necessarily non-volatile. Further most of the amino acids decompose before they melt.

3.2.3 General conclusions on protein structure

Surveying the 3-D structures of proteins that have been determined, some generalisations can be drawn about the distribution of charged and hydrophobic residues in these proteins:

'hydration shell'

- charged and polar side chains are generally found on the surface of the proteins where they interact with water molecules (the charges have a 'hydration shell' of water around them), thus solubilising the protein. It is also these surface charged groups that are responsible for the overall charge on the protein;

- hydrophobic side chains, not surprisingly, tend to be buried in the centre of the protein, away from the water. They often interact with one another via hydrophobic interactions and these interactions play an important part in maintaining the three dimensional structure of the protein. However, some hydrophobic side chains do appear on the surface, giving rise to occasional 'hydrophobic patches'.

We will see below how both these surface charges and hydrophobic patches are exploited in protein purification.

SAQ 3.2

You should now have a good mental image of the distribution of charged and hydrophobic groups in a typical globular protein. Which of the following statements generally apply to such a protein?

1) Hydrophobic side chains are predominantly on the outside of the molecule, with charged side chains buried inside the protein.

2) Because they are 'water hating' all hydrophobic side chains are to be found inside the protein structure.

3) The surface of globular proteins are highly positively charged.

4) Hydrophobic side chains are predominantly on the inside of the molecule, but some do exist on the surface.

5) There is no overall charge on the surface of proteins.

6) The nature of the overall charge on the surface of the protein depends on the pH of the protein's environment.

| SAQ 3.3 | Ovalbumin has a p*I* of 4.6. What will be the overall charge on ovalbumin at; 1) pH 3.5 and 2) pH 8.6? |

We will now look at individual methods available for purifying proteins.

3.3 Methods for purifying proteins

3.3.1 Introduction

The initial tissue extract from which one is attempting to purify a protein is necessarily a highly complex and concentrated mixture of proteins, probably together with nucleic acids, carbohydrates and inorganic salts. The general approach to protein purification is therefore to use initial preliminary fractionation methods, which, even though sometimes relatively crude, are able to cope with large amounts of protein and manage to remove a large part of the contaminating protein and other materials. This preliminary purification then allows one to move to the more sophisticated techniques, such as column chromatography, which have high resolutions but which tend to have low sample capacity and are fairly time consuming. These preliminary fractionation methods may include exploiting the thermal stability of the protein of interest, or may exploit the different distribution of charge and hydrophobic groups on proteins by using selective precipitation methods. The following therefore describes these preliminary fractionation methods followed by the more sophisticated column chromatography methods.

3.3.2 Thermal stability of proteins

The 3-D structure (tertiary structure) of globular proteins (for more details see the BIOTOL text, 'The Molecular Fabric of Cells'), is maintained by a number of forces, mainly hydrophobic interactions, hydrogen bonds and sometimes disulphide bridges. When we say that a protein is denatured we mean that these bonds have by some means been disrupted and the protein chain has unfolded to give the insoluble, denatured protein. One of the easiest ways to denature proteins in solution is to heat them. However, different proteins will denature at different temperatures depending on their different thermal stabilities; this, in turn, is a measure of the numbers of bonds holding the tertiary structure together. If the protein of interest is particularly heat-stable, then heating the extract to a temperature at which the protein is stable yet other proteins denature, can be a very useful preliminary step. For example, in the purification of the cytoplasmic enzyme aspartate aminotransferase (a relatively heat-stable enzyme) from a pig heart homogenate, the first step is to heat the extract to 75°C for 15 minutes.

thermal denaturation of proteins

A considerable precipitate of denatured protein appears during this treatment; this can be centrifuged down and discarded. The supernatant still contains the enzyme aspartate aminotransferase which has been considerably purified. A ten-fold increase in specific activity can be achieved in this way for this particular enzyme.

∏ Workers often include the substrate of the enzyme in the buffer solution when trying to purify an enzyme in this way. Can you think why?

The binding of the substrate at the active site of the enzyme will stabilise the structure of the enzyme and therefore enhance its stability (there can often be as many as a dozen hydrogen bonds between an enzyme and its substrate at an active site). Substrate is often included in buffer solutions, during purification protocols, to stabilise the enzyme.

A liver homogenate was shown to contain 50 mg cm^{-3} protein. Samples of the homogenate were assayed for the presence of three different enzymes. A, B and C and the results obtained were as follows:

enzyme A: 45 units cm^{-3};

enzyme B: 120 units cm^{-3};

enzyme C: 85 units cm^{-3}.

The homogenate was then heated to 80°C for 10 minutes, and precipitated protein removed by centrifugation. The protein content of the clear supernatant was shown to be 15 mg cm^{-3} protein. The three enzymes were again assayed and the results obtained were as follows:

enzyme A: 2 units cm^{-3};

enzyme B: 60 units cm^{-3};

enzyme C: 83 units cm^{-3}.

Would this heat treatment step be of any use in purifying any of the three enzymes? When assessing your reasons, try and introduce the terms 'specific activity' (ie units of enzyme per mg total protein) and 'yield'.

3.3.3 Fractional precipitation with ammonium sulphate

As the ammonium sulphate concentration of a crude mixture of proteins is slowly increased, more and more protein is seen to precipitate out of solution. Other salts may be used but ammonium sulphate is traditionally the most commonly used. By a process of trial and error, an appropriate concentration of ammonium sulphate can be found that will either:

- precipitate much of the contaminating proteins leaving the protein of interest in solution;

- precipitate the protein of interest, leaving much of the contaminating protein in solution.

It should be stressed here that when we precipitate the protein of interest we are precipitating native (ie active) protein that has become insoluble by aggregation; we have not denatured the protein. This precipitated protein can be recovered by centrifugation and the precipitate re-dissolved on adding buffer solution. In this method we are therefore fractionating proteins according to their different solubilities in ammonium sulphate solution. What is the principle behind this method?

As increasing amounts of ammonium sulphate are added to the protein solution, the salt dissolves because the ions become solvated by interaction with freely available water molecules. As more and more ammonium sulphate is added these freely available water molecules become scarce, so water molecules are preferentially pulled off from the hydrophobic patches on the protein surface. When the protein is dissolved in water the hydrophobic (literally water hating) patches are forced into contact with adjacent water molecules. It is not surprising therefore that these water molecules are easily given up to solvate the ions. The water molecules associated with the charged and polar groups on the surface of the protein on the other hand are bound by electrostatic interactions and are far less easily given up.

hydrophobic
interactions

As the ammonium sulphate concentration increases therefore the hydrophobic surfaces on the proteins are progressively exposed and proteins come together to form insoluble protein aggregates via hydrophobic interactions; ie protein - protein interactions become more important than protein - water interactions. Obviously those proteins with the largest number of hydrophobic patches aggregate first, whereas those with no hydrophobic surfaces probably will not precipitate even at high concentrations of ammonium sulphate. In a crude mixture, co-aggregation is obviously extensive; like molecules do not necessarily all stick together. Nevertheless, many proteins precipitate over a sufficiently narrow range of salt concentrations to make this a highly effective method of fractionation.

An added advantage of this method is that it concentrates the protein of interest. The protein of interest is usually present in small amounts in the initial tissue extract (you may be surprised to realise that water is the major contaminant in the early stages of purification!) but precipitation with ammonium sulphate allows you to re-dissolve the protein in a much reduced volume of buffer solution.

So far we have explained the salting out of proteins in a qualitative manner. In the next few paragraphs we will consider how the variation of the solubility of proteins with salt concentration can be explained quantitatively on the basis of physical chemical principles. This treatment has the advantage that by selecting the correct salt concentration it is possible to quantitatively precipitate out a single protein. It is necessary to introduce you to one or two mathematical equations; these will be quoted and not derived. However, it is important that you should be able to see that the equations are consistent with the experimental evidence, and appreciate their usefulness. Readers interested in the derivation of the equations will find this in standard textbooks of physical chemistry.

ionic strength

For this we need to understand the meaning of the ionic strength I, of a solution of electrolyte(s). The ionic strength, a concept first introduced by Lewis and Randall in 1921, is a measure of the intensity of an electric field due to all ions in solution. It is defined as half the sum of all the terms obtained by multiplying the concentration of each ionic species present in solution by the square of its charge; thus for fully dissociated electrolytes;

$$I = \frac{1}{2} \Sigma c_i z_i^2 \qquad \text{(E - 3.1)}$$

where c_i is the actual concentration of the ith ion and z_i its charge. For an electrolyte in which the concentrations of the ions are c_+ and c_- Equation (E - 3.1) becomes:

$$I = \frac{1}{2}(c_+ z_+^2 + c_- z_-^2) \qquad \text{(E - 3.2)}$$

Consider the ionic strength of 0.1 mol l^{-1} solution of sodium chloride, for this $c(Na^+) = c(Cl^-) = 0.1$ mol l^{-1} and $z_+ = z_- = 1$, hence;

$$I = \frac{1}{2}(0.1 \times 1^2 + 0.1 \times 1^2) = 0.1 \text{ mol } l^{-1}$$

Note that ionic strength has the units of concentration. For a 1,2 electrolyte (eg Na_2SO_4) or a 2,1 electrolyte (eg $CaCl_2$) of concentration c mol l^{-1} the ionic strength is equal to $3c$.

Try the following SAQ to see if you fully understand the principles involved.

<table>
<tr><td>

SAQ 3.5

</td><td>

Calculate the ionic strength of each of the following:

1)   0.5 mol l^{-1} solution of $NaNO_3$;

2)   0.5 mol l^{-1} solution of Na_2SO_4;

3)   1.5 mol l^{-1} solution of $(NH_4)_2SO_4$;

4)   solution of $NaNO_3$ (0.5 mol l^{-1}) and $(NH_4)_2SO_4$ (2.0 mol l^{-1}).

</td></tr>
</table>

We can now look at the way in which changing the ionic strength of a solution affects the solubility of sparingly soluble electrolytes in general and proteins in particular.

In solution of low ionic strength the solubility, S, of a sparingly soluble salt in a solution of an ionic strength, I, (the ionic strength is varied by the addition of an indifferent electrolyte) is related to the solubility S_o, at an ionic strength of 0 (an extrapolated value) by the equation:

$$\log \ S/S_o \ = \ 0.509 \left| \ z_+ z_- \right| \ I^{1/2} \qquad\qquad (E - 3.3)$$

'salting in' effect

where z_+ and z_- are the changes on the ions of the added electrolyte. This equation, which comes directly from the Debye-Hückel expression for the activity coefficient of a sparingly soluble salt, predicts an increase in the solubility arising from an increase in the ionic strength. This is known as the 'salting in' effect; this is seen in Figure 3.5.

As the ionic strength increases further Equation (E - 3.3), (a limiting equation applicable only at low I) is no longer valid and an additional term is included to give:

$$\log \ S/S_o \ = \ 0.509 \left| \ z_+ z_- \right| \ I^{1/2} - \ K'I \qquad\qquad (E - 3.4)$$

'salting out' effect

where K' is a positive constant whose value depends on the nature of the solute and the added electrolyte. The larger the solute molecule, in our case a protein, the greater is K', and at high ionic strengths $K'I >> 0.509 \left| \ z_+ z_- \right| \ I^{1/2}$. So now we can see from Equation (E - 3.4) that the ratio of solubilities (S/S_o) and hence the actual solubility, S, decreases with increasing ionic strength at high ionic strengths (note the negative sign in Equation (E - 3.4). This is known as the 'salting out' effect.

The change from increasing solubility at low ionic strengths to decreasing solubility at high ionic strengths is demonstrated in the experimental curves shown in Figure 3.5 for horse haemoglobin in a range of inorganic salt solutions.

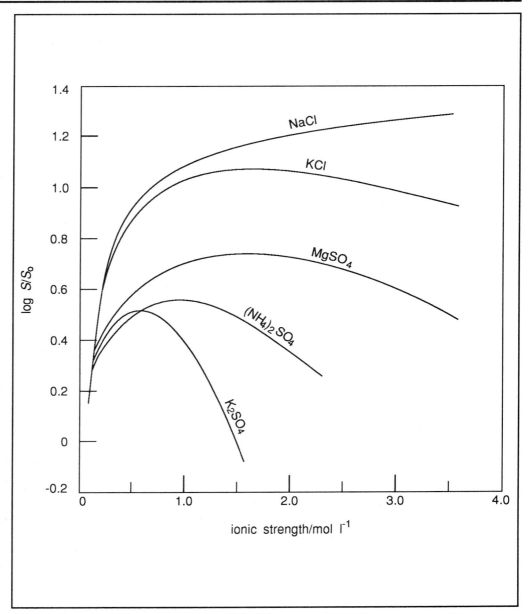

Figure 3.5 Plot of log S/S_0 against ionic strength for horse haemoglobin in the presence of various inorganic salts (stylised). Note that all the curves converge at $I = 0$ when $S = S_0$.

Figure 3.6 shows the salting out phenomenon for several proteins using ammonium sulphate. By carefully selecting the correct ionic strength, it is possible to quantitatively precipitate out a single protein. Since precipitation occurs over a small range of ionic strengths the separations are sharp.

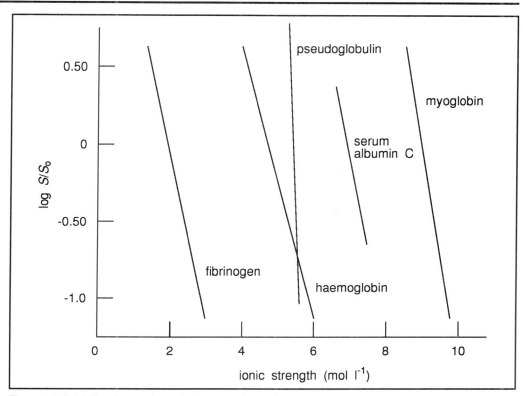

Figure 3.6 Solubility of several proteins in ammonium sulphate solutions.

SAQ 3.6

Solid ammonium sulphate was added to a tissue homogenate to give 35% saturation and the resultant precipitate was centrifuged down. The supernatant was collected and further ammonium sulphate was then added to the supernatant to give 45% saturation. The resultant precipitate was again collected by centrifugation. Continuing in this way a series of precipitates were collected at 35, 45, 55, 67 and 75% ammonium sulphate saturation. Each pellet was re-dissolved in buffer solution and assayed for both total protein, and the presence of the enzyme aspartate aminotransferase (AAT). The data obtained are tabulated below. Using the data, devise a purification step for AAT using ammonium sulphate fractionation.

% saturation of amonium sulphate	total protein in precipitate (mg)	total enzyme units in precipitate
35	200	0
45	450	15
55	500	270
67	400	20
75	150	5

3.3.4 Isoelectric precipitation

Many proteins show minimum solubility around their isoelectric point. By adjusting the pH of our extract to an appropriate value, it is possible to either 1) precipitate out the protein of interest while leaving many contaminating proteins in solution or 2) precipitate out some contaminating proteins while leaving the protein of interest in solution. Note again that the precipitated protein consists of aggregates of native proteins (it is not denatured) and can be re-dissolved by adjusting the pH. In a complex mixture there is of course co-precipitation of different proteins with similar properties; these aggregate to form the precipitate. Nevertheless this method, can constitute a useful preliminary fractionation.

What is the principle behind this method? Normally protein molecules in a mixture are kept apart by electrostatic repulsion. However, at its isoelectric point a protein has zero overall charge, thus minimising electrostatic repulsions. Like molecules can therefore get closer to each other allowing a positively charged region on one molecule to interact electrostatically with a negatively charged region on another. Since there are many charged regions on the surface of a given protein, large aggregates of protein can build up in this way leading to insoluble precipitates of native protein.

∏ Isoelectric precipitation is most successful when carried out at very low ionic strength (Figure 3.7). Can you think why this may be so?

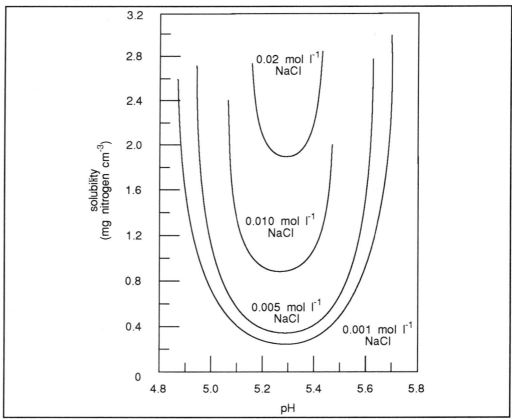

Figure 3.7 Solubility of β-lactoglobulin (pI = 5.3) as a function of pH at 4 different concentrations of sodium chloride.

In the presence of salts (eg NaCl), Na$^+$ ions will interact electrostatically with negatively-charged groups on the protein surface and effectively neutralise them, and likewise Cl$^-$ ions will interact with positively charged groups. This will therefore tend to mask the charged groups on the surface of the protein and therefore minimise the electrostatic interactions between protein molecules that we wish to encourage.

A variant of the last two methods involves the use of organic solvents, particularly acetone (propanone), methanol and ethanol, in attempts to precipitate certain proteins selectively from a mixture. These solvents often easily denature proteins, and the use of very low temperatures is advisable.

3.3.5 Gel filtration

This method most commonly used to separate proteins by size is called gel filtration chromatography, although various other names such as gel exclusion, gel permeation or molecular sieve chromatography, have been used for this method. (See also Chapter 1). The method involves passing the mixture of proteins through a column of small (μm) cross-linked porous beads of gel packed in an appropriate buffer solution. One of the most commonly used types of gel is Sephadex (beads of cross-linked dextran), but many others are used including Sepharose (cross-linked agarose) and Bio-Gel P (cross-linked acrylamide). The gel filtration media are sold in different grades depending on the fractionation range required. Sephadex G75, for example, fractionates proteins in the range of relative molecular mass (RMM) range of 3000 to 70 000.

Each gel filtration medium consists of a range of beads with slightly differing amounts of cross-linking and therefore slightly different pore sizes in the beads. The separation process depends on the different abilities of the various proteins to enter either some, all or none of the beads, which in turn relates to the size of this protein. For example, if a mixture of proteins is passed through a G75 column, all those proteins of RMM greater than 70 000 will be unable to enter any of the beads, since the pores in all the beads will be too small, and the proteins will therefore travel in the buffer solution surrounding the beads and elute first from the column.

void volume

The volume of solvent between the point of injection onto the column and the maximum of any eluted peak is known on the elution volume V_e. These large molecules (RMM 70 000) which are eluted from the column first are said to have been eluted in the void volume (V_o), also known as the totally excluded volume, since they were totally excluded from entering any beads. On the other hand, molecules of RMM less than 3000 are small enough to enter all the beads as they pass down the column and are therefore eluted much later in a volume equal to the total bead volume (V_t) also known as the 'totally included' volume since these molecules have entered all the beads. The purpose of running a G-75 column would be to separate proteins of RMM between 70 000 and 3000.

The following example shows how this is achieved.

∏ Let us assume that you have only three proteins, A, B and C in your mixture, and that the RMM are 60 000, 35 000 and 15 000 respectively. Will these separate on passing through a G-75 column, and if so, in what order will they elute from the column?

Yes, the proteins will separate, with A eluting first, followed by B, then C. At RMM = 60 000, protein A is close to the exclusion limit of this gel, but nevertheless there are a certain number of beads it can enter and therefore it will take somewhat longer to pass through the column than any totally excluded proteins, thus eluting a little after the void (V_o) volume. Protein C, on the other hand, being smallest will be able to travel through most beads, but not quite all, since there will be a certain proportion of beads where the pores are too small for the protein to enter, so the protein will elute a little before the totally included (V_t) volume. Protein B is intermediate, being able to enter roughly half the number of beads in the column, but being excluded from the other half. It will therefore elute between proteins A and C.

∏ Gel filtration columns tend to be very long columns, 1 metre or more in length. Why do you think this is?

Obviously the greater the column length the greater the number of beads the proteins can pass through/around and therefore the greater will be the separation of proteins. For example, a 10 cm column of Sephadex would be of little use as the difference in elution volume between V_o and V_t would be minimal, leaving little room for proteins to separate in between. However, the greater the bead volume the greater the volume difference between V_o and V_t and therefore the more likely proteins of different molecular masses are to separate.

SAQ 3.7

Sephadex G-50 fractionates proteins in the RMM range 1500 to 30 000. If a mixture of the following proteins was passed through the column, how many peaks of protein would you expect to elute from the column?

Protein	RMM
phosphorylase	92 500
transferrin	78 000
hexokinase	45 000
ovalbumin	43 000
lactate dehydrogenase	36 000
trypsin	23 800
soya bean trypsin inhibitor	20 100
lysozyme	14 400
insulin	5 800

3.3.6 Ion exchange chromatography

Ion exchange chromatography (see also Chapter 1) is the most commonly used method for protein purification. The method, which separates molecules according to charge, involves passing the mixture of proteins through a column of an ion exchange resin packed in an appropriate buffer solution. As you learnt in Chapter 1 an ion exchange resin is simply an inert support material, such as cellulose or Sephadex, to which have been attached positively or negatively charged groups. Two commonly used ion exchange resins, based on the inert support cellulose are shown in Figure 3.8.

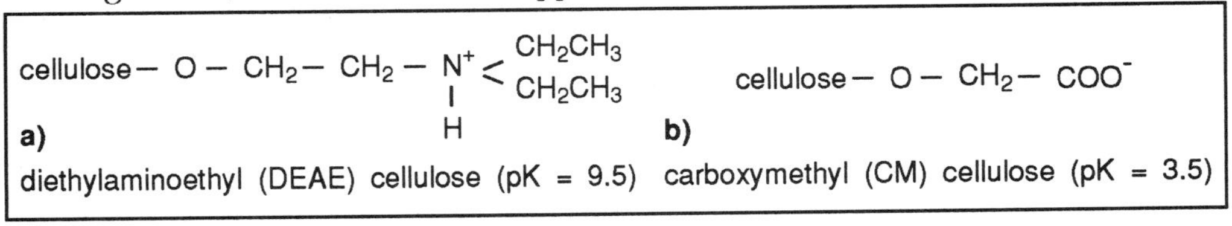

cellulose$-$ O $-$ CH$_2$$-$ CH$_2$$-$ N$^+$ $<$ $^{CH_2CH_3}_{CH_2CH_3}$ cellulose$-$ O $-$ CH$_2$$-$ COO$^-$

a) |
                                               H   b)
diethylaminoethyl (DEAE) cellulose (pK = 9.5)   carboxymethyl (CM) cellulose (pK = 3.5)

Figure 3.8 Commonly used ion exchange resins.

∏ What side groups are attached to cellulose in DEAE- and CM- cellulose? (Check Figure 3.8)

The answers are diethylaminoethyl- and carboxymethyl- groups.

DEAE-cellulose has a positive charge between pH 3 and 10 and therefore will bind negatively charged proteins in this pH range; DEAE-cellulose is therefore referred to as an anion exchanger. Conversely, CM-cellulose has a negative charge in the pH range 3 to 10 and therefore will bind positively charged proteins in this pH range; CM-cellulose is therefore referred to as a cation exchanger.

Π Consider a mixture of the following five proteins; pepsin, α-casein, insulin, chick haemoglobin and lysozyme. What will be the overall charge on each of these proteins at pH 5.6 (pI values are listed in Table 3.2)? Which of these proteins will bind to DEAE-cellulose if passed through a column of this material at pH 5.6?

Since DEAE will have a positive charge at this pH and pepsin, insulin and casein, being at a pH above their pI will have an overall negative charge, they should bind to the column. Haemoglobin and lysozyme will have an overall positive charge (the same as the column) and therefore will not bind to the column and will pass through.

Already, therefore, we have achieved a fractionation, some proteins being retained on the column, some having passed straight through. However, it would be to our advantage if we had chosen conditions (eg the pH at which the column was run) such that the protein of interest was one of those that bound to the column, as methods are available for eluting bound proteins in an organised manner.

The methods we can use rely on the fact that the proteins bound to the column are not bound to the column with equal strength. The following question will allow you to work out why this is so. (Before answering this question you might like to remind yourself of the principles of ionisation of amino acids and proteins covered at the beginning of this chapter (Section 3.2.1)).

Π In the example described above, indicate in decreasing order the relative strengths with which pepsin, insulin and casein are bound to the column.

Pepsin will have the strongest negative charge as it is furthest from its pI value and therefore will bind most strongly. α-casein will not bind so strongly as it is only 1.6 pH units above its pI, whereas insulin will bind least strongly as it is only 0.25 of a pH unit above its pI and will therefore only have a weak negative charge.

To elute these proteins from the column, we somehow need to break the ionic interactions between the protein and the ion exchange resin.

Can you think of any ways in which we can do this? There are two approaches we can use. Each involves applying some form of gradient in the buffer solution being pumped on the column. The first is to apply a pH gradient to the column, ie slowly increase or decrease the pH of the buffer solution being applied to the column.

The apparatus required is similar to that shown in Figure 1.2 for the formation of a continuous density column. In this instance, instead of having two solutions of different densities in the two vessels, we have a buffer solution in one and either acid or base in the other depending on whether we require a decreasing or increasing pH gradient.

Let us assume we apply a slowly decreasing pH gradient (starting at pH 8 to 9)to the column we have been discussing. What will this do to the charge on the proteins bound to the column? As the pH value decreases, so the overall negative charge on the protein will decrease until the pH of the buffer solution reaches the pI of the protein, at which point the protein will have no overall charge and will begin to elute from the column. In the example we have been discussing, insulin will be the first to be eluted as it will be the first to reach its pI, followed by α-casein and finally pepsin.

salt gradient | A more common approach is to apply an increasing gradient of sodium chloride. The chloride ions (Cl^-) will compete with the binding of the negatively charged proteins to the positively charged resin. In our example, insulin is most weakly bound to the resin and therefore can be displaced by a relatively low chloride ion concentration. As the chloride ion concentration increases, it will compete more strongly with the bound proteins and the α-casein will start to elute first and finally followed by the most strongly bound protein, pepsin.

SAQ 3.8

You are provided with a mixture of the following five proteins: ovalbumin, lysozyme, gelatin, fetuin and human haemoglobin. (p*I* values are given in Table 3.2).

Without having to apply any form of gradient to this column, identify an appropriate resin and pH of buffer solution for purifying 1) lysozyme; 2) fetuin.

3.3.7 Hydrophobic interaction chromatography

This form of chromatography involves binding proteins by hydrophobic interaction between their surface hydrophobic patches and a hydrophobic column matrix. The most commonly used column matrices are phenyl-sepharose (benzene rings attached to a sepharose support) and octyl-sepharose, (a short alkyl chain of 8 carbon atoms attached to a sepharose support). The hydrophobic interaction between protein and column is enhanced by the presence of salts which decrease the availability of water molecules in solution and remove water molecules from the hydrophobic regions of the proteins and resin, thus making them more accessible to interact. We have already covered exactly the same principle when discussing salt fractionation. Protein samples are therefore normally loaded in the presence of a high salt concentration which allows them to be retained on the column, and then eluted by gradually reducing the ionic strength by applying a salt gradient of decreasing ionic strength. As the ionic strength is slowly reduced, the more weakly hydrophobic proteins elute first followed by the more increasingly hydrophobic proteins. However, some highly hydrophobic proteins may not even be eluted in the total absence of salt. In this case it is necessary to add a small amount of water-miscible organic solvent such as propanol or ethylene glycol to the column buffer solution. This will compete with the protein for binding to the hydrophobic matrix and should elute the proteins.

3.3.8 Affinity chromatography (see also Section 1.3.4)

This method requires a specific molecule that binds to the protein of interest. For example, for an enzyme it may be the substrate or a specific inhibitor of the enzyme. Having identified the relevant specific molecule, this is then chemically linked (via a 'spacer arm') to an inert support matrix (eg agarose) to provide an affinity matrix material which can then be packed into a column. If a protein mixture is passed through this column, the enzyme of interest alone should bind while all other proteins should pass through the column unhindered. As you can see this should be a very selective way of purifying your protein, and indeed some proteins have been purified from crude homogenates in a single step using affinity chromatography.

spacer arm |

Π | You have achieved an excellent separation of a protein from contaminating proteins using an affinity column, but the protein is of no use to you stuck to the column! How do you think we can get the protein off?

The simplest and mildest approach is to pump a high concentration of ligand (in the above example the enzyme inhibitor or substrate) through the column. This will compete for binding to the enzyme with the much lower concentration of ligand on the column material, and in this way the enzyme (also known as the ligate) will be desorbed and eluted from the column.

ligand |

However, this approach is not always possible. It can, for example, not be used where an affinity column has been prepared using an antibody against the protein of interest. In this case we have to think of ways to break the electrostatic interaction between ligand and ligate. This can be done by changing the pH of the column buffer solution or by increasing the ionic strength of the buffer solution. The principle behind desorption in this case is exactly the same as described for ion exchange chromatography. Figure 3.9 shows schematically the principle behind affinity chromatography.

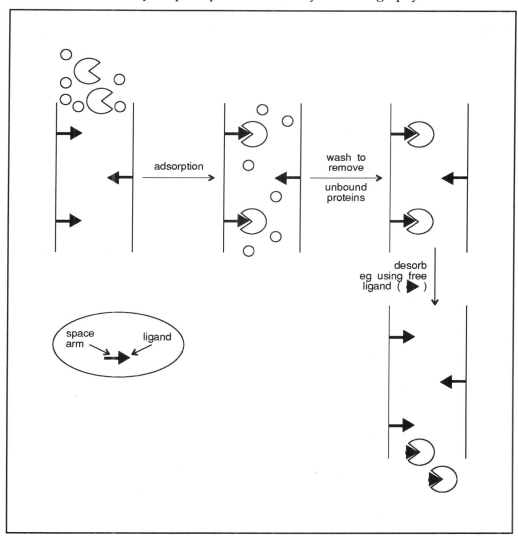

Figure 3.9 Schematic representation of enzyme purification using affinity chromatography.

3.4 When to stop?

It is all very well having all these purification methods available, but how do you know when to stop?! In other words, how do you monitor the purification process and what will ultimately convince you that your protein is pure?

∏   What tests would satisfy you that your protein was pure?

When purifying enzymes, the specific activity of an enzyme should increase at each step, while for a purified enzyme the specific activity cannot increase. A maximum value for specific activity is therefore a good indication that the enzyme is pure.

However, there are more direct ways of observing the purity of a protein that are routinely applied.

- The presence of a single band following gel-electrophoresis usually on a native or denaturing acrylamide gel (see Section 1.3.4) is highly indicative of purity. All initial crude extracts show a highly complex pattern of proteins on electrophoresis. However, as purification of the protein proceeds, this pattern simplifies until finally, when there is a single protein, only a single band, should be observed.

- There are methods that will measure the number of different N-terminal amino acids in a protein solution. A pure protein should only show the presence of a single N-terminal amino acid.

- If a protein shows a single band on gel-electrophoresis and a single N-terminal amino acid, one can be fairly certain the protein is pure.

∏   Can you think of any occasion when more than one N-terminal amino acid is detected and we still have a pure protein?

This would occur when the protein consisted of more than one protein chain, eg haemoglobin which has two α and two β-chains each of the two chains having a different N-terminal amino acid. Pure haemoglobin therefore shows the presence of two different N-terminal amino acids. This problem is addressed further in Chapter 5.

SAQ 3.9

Which of the following methods can be used to purify proteins?

Separation according to: 1) size; 2) colour; 3) charge; 4) solubility; 5) density; 6) stability; 7) volatility; 8) viscosity.

SAQ 3.10

Which of the following statements are correct?

In ion-exchange chromatography, proteins bound to the column can be eluted by:

1) change of pH;

2) change of temperature;

3) change of pressure;

4) by applying a salt gradient;

5) by increasing the flow rate of the buffer solution.

SAQ 3.11

Which of the following statements are correct?

Ammonium sulphate precipitation is often used because:

1) it is inexpensive;

2) it is very specific;

3) it only precipitates enzymes;

4) it can be carried out on a large scale;

5) it separates on the basis of charge.

SAQ 3.12

Which of the following statements are correct?

Affinity chromatography separates proteins on the basis of:

1) the sum total of the charge on the protein;

2) the overall size of the protein;

3) the hydrophobicity of the protein;

4) a support ligand which has a shape and charge complementary to that on a region of the protein;

5) the isoelectric point of the protein.

SAQ 3.13

Which of the following statements are correct?

The isoelectric point of a protein is:

1) the pH at which the protein has maximum positive charge;

2) the pH at which all ionisable groups on the proteins are uncharged;

3) the pH at which the protein has maximum negative charge;

4) the pH at which the protein has an overall zero charge.

Summary and objectives

The principles and practice of several methods used to purify proteins are described and their relative advantages and disadvantages are discussed. As an aid to understanding these methods, a general review of protein structures is included; this considers the properties of the different amino acids which constitute a protein and the consequences of the 3-dimensional structure of proteins. Particular attention has been paid to the hydrophilic and hydrophobic side chains.

On completion of this chapter you should be able to:

- explain the meaning of the isoelectric point of amino acids and proteins and describe how the charge on these varies with pH;

- appreciate the importance of the differences between the hydrophilic and hydrophobic side chains of amino acids;

- describe how the thermal stability of proteins can be used as a method for their purification and/or isolation;

- calculate the ionic strength of an electrolyte solution and explain qualitatively and quantitatively the process of 'salting in' and 'salting out';

- give an account of the principles underlying isoelectric precipitation, gel filtration, ion exchange chromatography, hydrophobic interaction chromatography and affinity chromatography;

- establish the purity of a protein, including enzymes, at various stages in the purification process;

- select suitable physical methods to decide if a protein is pure.

The purification and determination of the amino acid sequence of peptides

The purification and determination of the amino acid sequence of peptides

4.1 Introduction

The first question we should ask is what is the difference between a peptide and a protein?

∏   Describe the difference between a peptide and a protein?

It is simply a matter of size. Both are polypeptides - in other words they are both made up of chains of amino acids. If it is a short chain of amino acids, say less that fifty, we call it a peptide; if greater, we call it a protein. In practice, many peptides found in biological systems consist of very few amino acids, often less than ten, whereas proteins consist of at least one hundred amino acids, and often many hundreds of amino acids linked together.

Peptides and proteins are therefore at opposite ends of the polypeptide spectrum. Many peptides found in biological systems are of great importance because of their biological activity. Can you think of any peptides that you know of, and if so, do you know what their activities are? Table 4.1 lists some of the better studied ones.

Name	Number of amino acids	Function
angiotensin II	8	vasoconstriction, resulting in increased blood pressure
bradykinin	9	contraction of smooth muscle
corticotrophin (also known as ACTH)	39	promotes growth of adrenal cortex and stimulates it to synthesise steroid hormones
enkephalins	5	control of pain
gastrin	17	regulates secretions in the alimentary tract
glucagon	29	stimulates glycogen breakdown in the liver
gramicidin A	15	antibiotic
insulin	51	stimulates glucose uptake by cells and glycogen synthesis
oxytocin	9	contraction of smooth muscle
secretin	27	regulates secretions in the alimentary tract
vasopressin (antidiuretic hormone)	9	control of water balance by stimulating water resorption

Table 4.1 Some well known peptides.

insulin

You will notice in Table 4.1 that insulin contains 51 amino acids. This shows the arbitrariness of deciding that 50 is the cut off point for differentiating peptides and proteins, since most people refer to insulin as a peptide. However, you will notice that all the others are considerably smaller and unambiguously can be called peptides.

With the exception of Gramicidin A, which is an antibiotic, all the other peptides are mammalian hormones with essential biological activities. It is not surprising therefore that a lot of work continues to be put into purifying and characterising peptides (which includes working out the amino acid sequence) from biological tissues and fluids. In Chapter 5 you will come across another reason for wanting to purify peptides. This is when you cleave a protein into smaller peptides for sequence determination.

Because of their potent biological activity, peptide hormones tend to be present in very low levels in animal tissue, and therefore their purification and characterisation is a very challenging task. To give you an idea of the nature of the problems that can exist, Schlesinger and co-workers purified 18 µg of pig corticotrophin releasing factor (pCRF), which was sufficient to allow them to determine the complete amino acid sequence of 41 residues. However, to obtain their eighteen microgrammes, they used as their starting material the hypothalamus from 479 000 pigs! We will now examine the methods available for purifying peptides.

4.2 Peptide purification

4.2.1 Separation techniques

We usually have to purify peptides from a cell or tissue homogenate. Our first task therefore is to 'extract' the peptide fraction from this homogenate. As well as peptides, the homogenate will contain, amongst other components, proteins and nucleic acids.

∏ Can you think of a simple way in which we can separate the proteins and nucleic acids from the peptides?

size

The proteins and nucleic acids will be considerably larger than the peptides. We can therefore separate these on the basis of size using gel filtration (Sections 1.3.4 and 3.3.5) for example, we could use Sephadex G-25. All molecules of relative molecular mass (RMM) greater than 5000 (in other words the proteins and nucleic acids) will pass through in the excluded volume, whereas the smaller peptides will come off the column later, thus separated from the protein and nucleic acids. We now have a mixture of peptides and other small molecules and can concentrate on separating out the peptide of interest.

Because of their small size, peptides tend to exist in solution as fairly random chains, certainly they do not have the complex, folded three-dimensional structure of proteins. It is not possible, therefore, to denature peptides (as there is no structure to denature) and consequently peptides tend to remain soluble at most pH values. Methods which we use to fractionate proteins, that rely on differences in solubility or stability, such as ammonium sulphate precipitation, isoelectric precipitation, heat stability (Section 3.3) cannot therefore be used for peptides. Also of course the extremely low concentrations and amounts of most peptides means that aggregation and precipitation methods would be of no use anyway. However, like proteins, peptides do differ in other characteristics such as charge, size and hydrophobicity.

☐   What were the methods we used to separate proteins based on 1) size and 2) charge?

They were 1) gel filtration and 2) ion exchange chromatography and electrophoresis. (If you have forgotten details of these refer back to Sections 1.3.4, 3.3.5 and 3.3.6).

These methods have been used in the past to fractionate mixtures of peptides. So indeed has the separation of peptides by either chromatography or electrophoresis on filter paper. However, it would be misleading to suggest that these methods are regularly used nowadays. This is because one technique alone has proved so successful in peptide purification that it tends to be used to the general exclusion of all other methods. The method is reverse phase HPLC.

4.2.2 Reverse phase HPLC

HPLC stands for 'high performance liquid chromatography' (Section 1.3.2). The reason for the phrase 'high performance' will be explained later. 'Reverse phase' (RPC) refers to the fact that we are separating molecules based on their distribution between a polar mobile phase (the liquid you are passing through the column) and the column material which is an organic (non-polar) phase attached to an insoluble support.

☐   Why do you think this form of chromatography is called 'reverse phase' chromatography? It will help if you can decide what 'normal phase' chromatography is.

In normal phase chromatography (for example, when separating samples on a TLC plate), the stationary phase is polar (silica or cellulose) and the mobile phase usually non-polar. In RPC these are reversed. The stationary phase is non-polar (the C_{18} material) and the solvent is polar.

reverse phase chrom- atography
    Column materials used for reverse phase chromatography are often aliphatic chains (the 'non-polar' phase) chemically bonded to insoluble silica beads. These aliphatic chains are simply chains of $-CH_2-$ groups. The most commonly used groups for peptide purification are the so called C_{18} columns where the alkyl chain is 18 carbon atoms long. These columns are also commonly referred to as ODS (octadecyl silyl) columns. A large number of these chains are attached to each small bead resulting in a high density of hydrophobic alkyl groups attached to the silica packing.

To carry out reverse phase chromatography, the mixture to be fractionated is loaded onto the RPC column and the peptides bind to the column by hydrophobic interaction between hydrophobic groups on the peptides and the C_{18} chains. The less polar a peptide is, the stronger its interaction with the hydrophobic stationary phase will be. These bound molecules are then eluted sequentially from the column by applying a gradually increasing concentration of a water-miscible organic solvent, normally acetonitrile (methyl cyanide) or methanol.

This organic solvent will compete with the peptides for binding sites on the C_{18} resin. Obviously the least hydrophobic peptides will be eluted first, but as the organic solvent concentration increases, so eventually even the most hydrophobic peptides should be eluted. This method gives excellent separation of peptides. Peptides eluting from the column are monitored by passing the effluent stream through a flow cell in a UV spectrophotometer and then collected in a fraction collector.

∏   What wavelength would you use to monitor the eluting peptides: 280 or 220 nm?

At 280 nm we would only detect peptides containing tyrosine and tryptophan. Many peptides would not contain these amino acids, and therefore would not be detected. However, by monitoring at 220 nm we are detecting the peptide bond and therefore will record all peptides eluting from the column (Sections 1.5.2 and 2.2).

A typical separation of a peptide from a tryptic digest of a protein using reverse phase chromatography on a C_{18} column is shown in Figure 4.1.

In the earlier quoted example of Schlesinger's work on pCRF, following initial separation of proteins and nucleic acids from peptides using Sephadex G-25 gel filtration, the pCRF was purified using three chromatographic runs on reverse phase columns, using slightly different conditions each time.

Hopefully the above theory sounds reasonable to you. However, it does seem to rely on the presence of hydrophobic amino acids in the peptide to allow the peptide to bind to the column. Surely many peptides will have no hydrophobic amino acids? Indeed, many will be quite the opposite. There will be some highly charged peptides that are very hydrophilic and therefore will certainly not want to bind to a hydrophobic surface. This is indeed true, but we can still separate these by reverse phase chromatography. This is achieved by exploiting the principle of ion-pairing. RPC of peptides is carried out at low pH (~pH 2) and trifluoracetic (CF_3COOH) is usually used to achieve this pH.

ion paring

The negatively charged trifluoracetate group (CF_3COO^-) will bind to any positively charged groups in the peptide (this is the 'ion-pairing') thus masking this charge. Also the CF_3-group is hydrophobic in character, so not only have we removed a charged group from the peptide, we have also converted it into a group with a hydrophobic character.

∏   What positively charged groups might we find on a peptide?

The N-terminal amino group of the peptide will exist as NH_3^+. Also, the side chains of any lysine or arginine residues in the peptide will also have a positive charge (Table 3.1).

Ion-pairing in this way therefore seems to cope with positively charged groups in our peptides and considerably increases the hydrophobic nature of the peptide. There are also negatively charged carboxyl groups in peptides which need to be considered.

∏   Can you remember where these carboxyl groups come from?

There will be one at the C-terminus of the peptide, and one on each of the side chains of any aspartic or glutamic residues present.

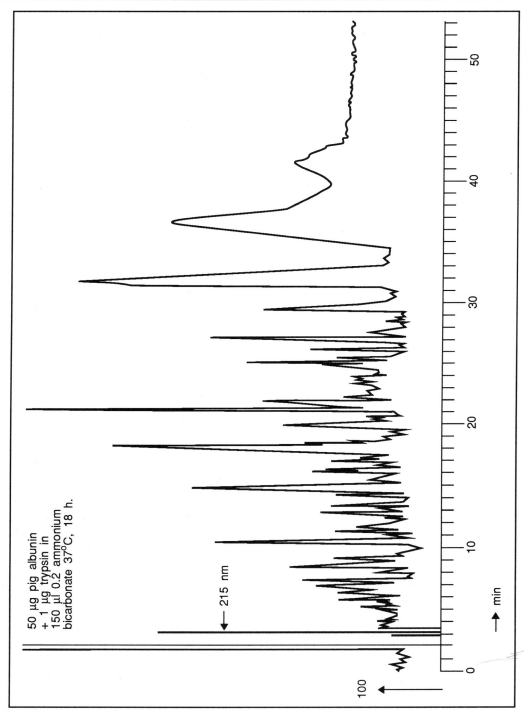

Figure 4.1 Typical separation of peptides from a tryptic digest of a protein using reverse phase chromatography. Kindly provided by BJ Smith, Celltech Ltd, Slough. Key: HPLC = Hewlett Packard 1090: column = sorbax C8: buffer A = H$_2$O; B = acetonitrite; C = 0.5% TFA in H$_2$O: Gradient = t = 0 min:88% A; 2%B; 10% C; t = 40 min: 30% A; 60% B; 10%C: t = 60 min: 30%A; 68%B; 10%C, (then washed and

$\prod$ What will happen to these negatively charged groups under the conditions we are using, ie at pH 2?

They will tend to be protonated, thus losing most of their negative charge. In the following equation the equilibrium will be to the right.

$$-COO^- + H^+ \rightleftharpoons -COOH$$

Therefore, a further advantage of working at low pH is that we are effectively removing most of the negatively charged groups. This reduction in the polarity of the peptide will further enhance its interaction with the hydrophobic stationary phase. By carrying out RPC of peptides in trifluoracetic acid (usually 0.1%) we therefore convert hydrophilic molecules into essentially hydrophobic ones which can now be separated by reverse phase chromatography. Figure 4.2 shows an example of the effect of putting a hydrophilic peptide into a low pH environment formed with trifluoracetic acid.

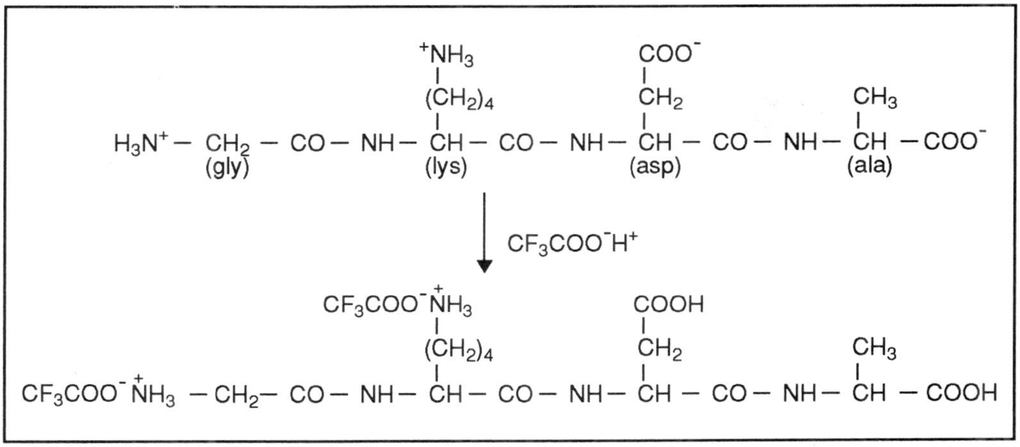

Figure 4.2 Diagram showing the effect of dissolving a hydrophilic peptide in 0.1% trifluoracetic acid.

The above discussion has centred around reducing the polarity of peptides to ensure that they bind to a reverse phase column. However, one can sometimes experience the reverse problem, where peptides are already extremely hydrophobic and therefore bind strongly to the column and are therefore difficult to elute.

$\prod$ In this case what can we do to reduce the strength of binding of these peptides to the column?

There are two approaches, either or both of which could be used. Firstly, one can use a less hydrophobic column, perhaps a C_8 or even a C_4 column. These columns have shorter alkyl chains and therefore result in a lower density of alkyl groups and are therefore less hydrophobic. Binding of peptides should therefore be weaker.

Secondly, one could carry out ion-pairing with phosphoric acid, rather than (trifluoracetic acid) TFA. The phosphate ion is hydrophilic in nature, and therefore this hydrophilic ion pairing (ion-pairing with TFA is hydrophobic ion-pairing) will maintain the polar groups on the peptide and hence reduce the interaction with the hydrophobic stationary phase.

Π Peptide mixtures that we want to separate often contain quite a lot of salts (such as components of buffer solutions). For example, if we have digested a protein with an enzyme to produce a peptide mixture, the buffer solution used for this reaction, for example Tris, will be present. When we adjust the pH of our mixture to pH 2 and load it onto the RPC column, what will happen to contaminating salts?

They will pass straight through the column, since there is no reason why charged ions should bind to the hydrophobic support. RPC is, therefore, also a useful way of 'desalting' peptides. When the peptides are eluted from the column, they can be dried under vacuum where both the trifluoracetic acid (which is highly volatile) and the organic solvent will evaporate leaving the pure peptide. Desalting peptides to leave them free of salt is very important since salts would otherwise interfere with sequencing studies that are likely to follow peptide purification.

4.2.2 Why is HPLC so successful?

HPLC is not only used to purify peptides. It is the method of choice in nearly all areas of the biological sciences where one is separating or purifying small organic molecules, for example when studying drug metabolites, purifying carbohydrates, nucleic acids, lipids. It is hard to imagine any research laboratory nowadays without an HPLC machine.

principles of HPLC

Why therefore is HPLC so successful? It is important to stress that the principle behind the separation on an HPLC column is not different to that used in traditional column chromatography. Separation on an HPLC column is achieved by the same techniques as conventional column and gel filtration or ion exchange chromatography. However, it is the improved design of the packing material, achieved by the manufacturers, that is responsible for the exceptional resolution of an HPLC column. We need therefore to explain why an HPLC column of only, say, 15 cm length, has such superior resolution power compared to say a more conventional column which is very much longer and contains considerably more packing material. In other words, what is responsible for this 'high performance'?

Let us take as an example, a gel filtration column. In a commonly used support such as Sephadex, the beads are not of uniform size. Therefore when the column is packed, there will be open regions of liquid between the beads (Figure 4.3 a)).

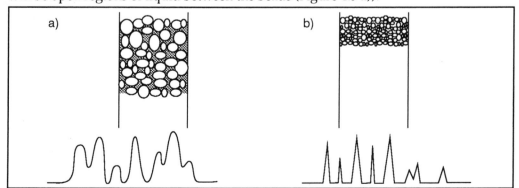

Figure 4.3 Diagram showing: a) column packing, and associated 'dead spaces' with conventional column packing, together with a typical elution profile. Note the peak broadening caused by diffusion in the 'dead spaces'; b) column packing with HPLC column material, and associated elution profile. Note that the peaks are much sharper, due to the lack of space for diffusion to occur between the beads. Note also that in an ideal column, the beads should be of the same size. Manufacturers strive to produce uniform beads but, in practice there is inevitably some size distribution.

As the sample moves down the column, separation occurs by interaction of the different species with the beads. However, when the species are in the liquid between the beads, diffusion can take place and this leads to band broadening. Separated fractions are therefore eluted as fairly broad peaks as shown in Figure 4.3 a). However, in an HPLC column, all the particles are of nearly identical small size, which achieves close packing of spheres, with minimal free space between the beads. This leads to far less diffusion and hence much sharper peaks (Figure 4.3 b)).

Secondly, it should be apparent that the amount of separation depends on the number of interactions that occur between the molecules being separated and the column material. The greater the surface area of column material available, the greater the separation that will be achieved. Increasing the surface area available is achieved by using extremely small beads (for example, 5-10 μm in diameter).

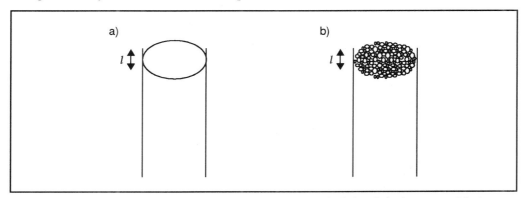

Figure 4.4 Diagram showing a) hypothetical situation where just a single bead of column material takes up the space over a column; b) the same space, packed with HPLC column material. A mixture will come into contact with a much greater surface area of bead when moving a distance *l* in column b) compared to column a).

The smaller the beads are, the greater the total surface area of beads one can get into a given space. Figure 4.4 a) shows a highly over-exaggerated situation where just one bead occupies a given length, *l*, in a column. If the same column length was to be packed with much smaller beads (Figure 4.4 b)), it should be clear that in moving the distance (*l*) through the column the molecules to be separated will come into contact with a vastly greater surface area of column material, and the separation will therefore be considerably greater.

SAQ 4.1

You are required to prepare a chromatography column, the total volume of packing material is 10 cm^3 (excluding any void space between the beads of the material). You are provided with spherical beads of diameter; 1) 0.2 mm and 2) 5 μm. Calculate the total area available for the two different types of bead and hence decide which material would be preferred for use in the column.

It is the manufacturers' achievement of producing, in a reproducible manner, small beads, of uniform size, that has led to the development of HPLC. However, because the packing material is so fine, it is difficult to pump solvent through the column. For this reason high pressure pumps have to be used to achieve any flow at all through the HPLC column. This is why HPLC is sometimes incorrectly referred to as 'high pressure liquid chromatography'. It must be stressed that the high pressure has absolutely nothing to do with improving resolution on the column; it is simply necessary to get solvents to flow through the column. Pressures up to 2000 psi are used, and for this reason HPLC columns are packed in stainless steel tubes, using stainless steel tubing with pressure joints to link pumps, solvent and column. If conventional glass columns and plastic tubing were used they would of course explode under the high pressure.

The need for high pressure introduced one further problem for the manufacturer. They could not use traditional support materials for the column packing such as cross-linked dextran chains (eg Sephadex). These beads are highly compressible, and would distort, flatten, and block the column under such high pressures. This is why most HPLC column packings are made from silica beads, since these are resistant to distortion under high pressure.

It is therefore the development of small, uniform-sized column materials, appropriately derivatised (for example, by the addition of alkyl chains to the surface of silica beads for RPC) that has resulted in HPLC being the method of choice for many purification and analytical purposes. A more mathematical approach to calculating the resolving power of HPLC and other chromatographic procedures is given in the BIOTOL text, 'Product Recovery in Bioprocess Technology'.

4.2.4 Purity of the peptides

The question now arises is how do we know when a peptide is pure?

A single sharp peak eluted from a reverse phase column is a good indication that a peptide is pure. However, one may be unlucky and have two peptides with similar characteristics eluting in the same place. The best test for purity is to show the presence of only a single N-terminal amino acid in the sample.

N-terminal amino acid

If we have a pure peptide, every molecule in the solution will have the same N-terminal amino acid. Therefore, if we have a test for detecting N-terminal amino acids, we should detect only one. However, if there is more than one peptide present, say there are three, then we would most probably detect three N-terminal amino acids and would know the peptide was not pure.

dansyl method

The method most commonly used for determining N-terminal amino acids is known as the dansyl method. A small sample of the peptide is reacted with dimethylaminonapthalene-5-sulphonyl chloride (dansyl chloride) for about 1 hour. The dansyl group reacts with free amino groups and will therefore react with the N-terminal amino acid in the peptide. Following this incubation, the sample is dried, dissolved in 6 mol l^{-1} HC1 and heated at 105°C for about 18 hours.

∏   What do you think this treatment will do to the peptide?

These are the standard conditions routinely used for hydrolysing peptide bonds in proteins or peptides to give a mixture of free amino acids. The peptide will therefore be broken down to its component amino acids.

Although all the peptide bonds will be cleaved by this treatment, the linkage between the dansyl group and the N-terminal amino acid is resistant to hydrolysis under these conditions. Following hydrolysis therefore we have a mixture of amino acids, but one of them, the N-terminal amino acid (the one we are trying to identify), has been 'tagged' by this dansyl group, and thus become a dansyl amino acid.

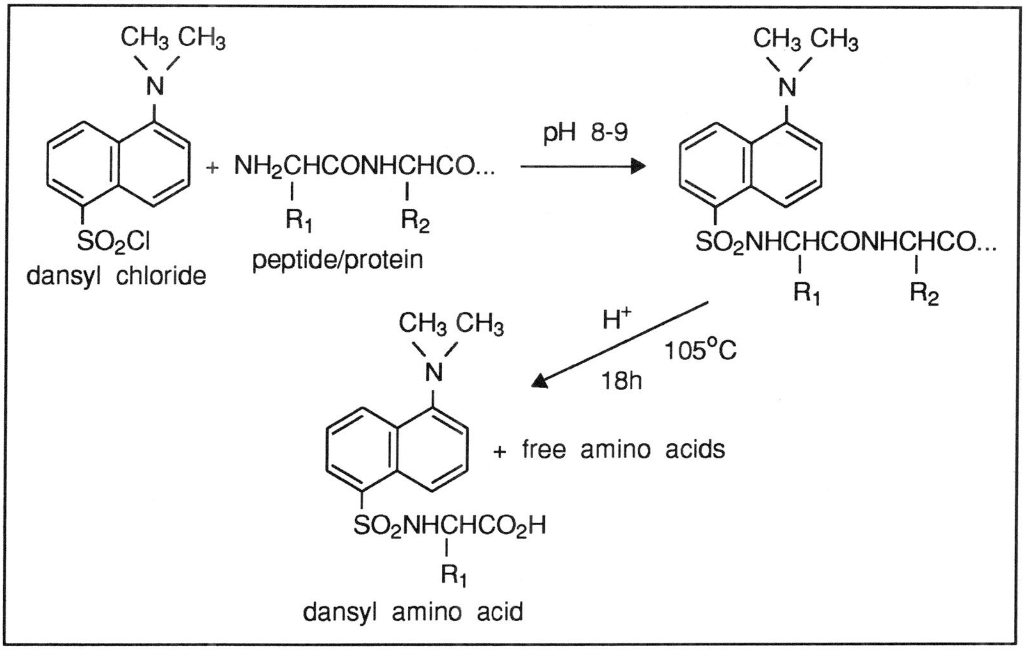

Figure 4.5 The dansyl method for N-terminal amino acid analysis

Our task is therefore to identify this labelled amino acid. This is relatively easy since dansyl amino acids are fluorescent under UV radiation, whereas the unlabelled amino acids are not. A simple 2D thin layer chromatography system is used which takes about 30 minutes to run. When UV radiation is shone on the chromatogram the dansyl amino acid shows up as a green fluorescent spot. Although all the other amino acids present in the peptide are also on the plate they are not seen since they are not fluorescent. If we see a single fluorescent spot on the plate we know that the peptide is pure. In addition, by observing the position of the spot on the plate and comparing this with the position of known dansyl amino acids run at the same time one can determine which dansyl amino acid it is. For example, it might be dansyl alanine, so that we know the N-terminal amino acid of the peptide is alanine.

4.3 Peptide sequence determination

The ultimate characterisation of a peptide is the determination of its amino acid sequence. Because of their relatively small size, the sequencing of peptides is not a particularly daunting task. You will see in Chapter 5 that, not surprisingly, much more work is involved in sequencing a protein.

4.3.1 Chemical method

Edman degradation

In 1953, Pehr Edman introduced a chemical method for determining the order of amino acids in polypeptides, and the method has become known as the 'Edman degradation'. Despite tremendous advances in protein chemistry technology in recent years, the Edman degradation remains, nearly 40 years after its introduction, the main method of choice for sequencing proteins. In this method amino acids are removed one at a time from the N-terminus and each identified. In this way the order of the amino acids from the N-terminus is determined.

The removal and identification of the N-terminal amino acid from a polypeptide by the Edman degradation involves three simple chemical steps. We refer to this as carrying out one cycle of the Edman degradation. These steps are shown in Figure 4.6.

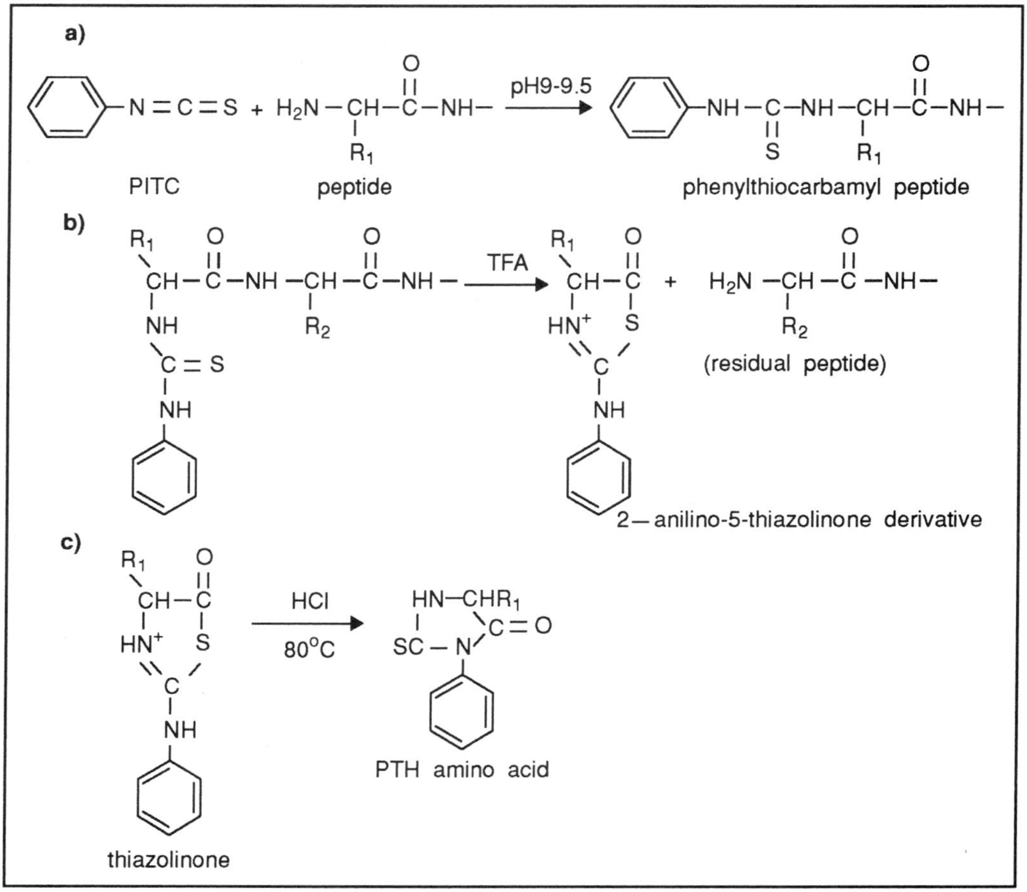

Figure 4.6 The Edman degradation. Key: a) the coupling reaction; b) the cleavage reaction; c) the conversion reaction. (PITC = phenylisothiocyanate; TFA - trifluoracetic acid; PTH - phenylthiohydantoin derivative).

The first step is the coupling reaction. The peptide is incubated with phenylisothiocyanate (PITC). This reacts chemically with the amino group of N-terminal amino acids to give what is know as the phenylthiocarbamyl (PTC) derivative of the peptide.

The second step is the cleavage reaction. The PTC peptide is dried under vacuum and then dissolved in an anhydrous acid (usually trifluoracetic acid, TFA) and incubated at $50°C$. Under these conditions the peptide bond between the first and second amino acid is cleaved. The first amino acid in the peptide chain is therefore released as a derivative known as a thiazolinone. The actual incubation times depend on whether or not a manual or automated system is used.

The first amino acid has now been cleaved from the peptide and has a phenyl group attached to it. It is therefore relatively hydrophobic and can be extracted using an organic solvent, for example, benzene. The rest of the peptide remains in solution since it is hydrophilic and not soluble in benzene. All we now need to do is identify this extracted thiazolinone. However, thiazolinones are not particularly stable, so rather than identify them directly, we must carry out the third and final step known as the conversion step. The extracted thiazolinone is heated in 1 mol l^{-1} HCl at 80° for 10 minutes, under which conditions it is isomerised to the more stable phenylthiohydantoin (PTH) derivative. Our task is therefore to identify this PTH derivative. Because of the phenyl group present in each PTH amino acid they are relatively hydrophobic and can therefore be separated and identified by reverse phase chromatography.

Having identified the first amino acid, we can now return to the rest of the peptide and carry out another cycle of the Edman degradation and remove the second amino acid, which of course is now at the N-terminus of the peptide, and similarly identify this second amino acid as its PTH derivative. This process of repeating the cycle and removing and identifying one amino acid at a time continues until the sequence of the peptide is determined.

This probably makes peptide sequencing sound very simple, but there are problems. The Edman degradation is not 100% efficient. When carrying out sequencing in a test-tube (so-called 'manual sequencing') the reaction is only about 93-95% efficient). Therefore, after say 10 cycles of manual sequencing, many peptide molecules will have had 10 residues removed, but some will only have had 9 residues removed, a few only 8. In other words, the N-terminal sequence is becoming staggered. The more cycles one carries out, the greater the stagger and the more difficult it is to identify the newly liberated N-terminal amino acid as the 'main' sequence. Table 4.2 shows the importance of repetitive yield in the amount of sequence data one can obtain when sequencing using the Edman degradation.

Repetitive yield/%	Residues of sequence
95	15
96	25
97	40
98	60

Table 4.2 The relationship between repetitive yields in the Edman degradation and the average number of residues of sequence successfully obtained.

When carrying out manual sequencing, therefore, even on a peptide of only 15-20 residues, some difficulty can be experienced in unambiguously confirming the sequence of the last few residues of the peptide. It would help enormously if we had another method where we could determine the sequence starting from the C-terminus. Unfortunately there is no chemical method analogous to the Edman degradation for sequentially removing amino acids one at a time from the C-terminus. However, limited information can be obtained using enzymes.

4.3.2 Enzymic method

C-terminus of polypeptides

The enzyme carboxypeptidase Y (from yeast) cleaves amino acids one at a time from the C-terminus of polypeptides. Therefore if we incubate our peptide with the enzyme and remove aliquots at timed intervals to determine the free amino acids present in solution, by plotting a graph of amount of amino acid released against time, we can certainly deduce the last two or three amino acids in the sequence. With extended cleavage times, the pattern of release of amino acids becomes quite complex and difficult to interpret accurately. However, having identified the last two or three amino acids it is possible to use this information to confirm the results of the latter stages of the N-terminal sequence data. Figure 4.7 shows a typical result from a carboxypeptidase experiment.

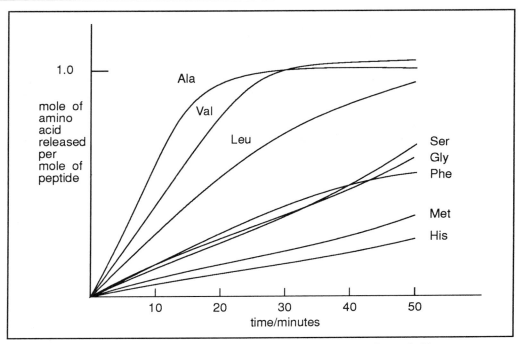

Figure 4.7 Graph showing release of free amino acids with time for a peptide digested with carboxypeptidase Y.

⊓ What can you deduce about the C-terminal sequence of the peptide being studied in Figure 4.7?

Alanine is obviously the first amino acid to be released and is therefore the C-terminal amino acid. The next to be released is clearly valine, and the next probably leucine. After this the pattern of release becomes complicated and no more further information can be deduced. The peptide therefore ends: Leu-Val-Ala.

Manual sequencing is reasonably successful for peptides up to 15 to 20 residues. For larger peptides the use of an automated protein/peptide sequencer is required. These are discussed further in Chapter 5, but basically because of their design, the repetitive yield can approach 98-99%, thus allowing much longer peptides to be successfully sequenced. It should be noted, however, that the automated sequencer is a very expensive piece of apparatus. Many workers do not have access to this and must rely on 'manual' sequencing of their peptides.

4.3.3 Mass spectrometry for sequencing peptides

The use of the Edman degradation has been, and continues to be, the method of choice for sequencing peptides. However, there is an alternative approach, the use of mass spectrometry (MS), which, because of new instrumentation and improved techniques, is becoming a highly attractive alternative approach to peptide sequencing. It is beyond the scope of this chapter to describe the analysis of peptides by mass spectrometry in any detail, but you should at least be aware of these developments.

The theory and practice of mass spectrometry are described in the BIOTOL text, 'Techniques used in Bioproduct Analysis'. To analyse samples by mass spectrometry they must be in the vapour phase, that is they must be volatile. Peptides are essentially ionic compounds and therefore initial attempts to analyse peptides by MS required extensive chemical degradation and derivatisation to get them into the vapour phase. Despite these limitations, methodologies have been developed that allow sequence data to be obtained for peptides up to about 15 residues in length.

Recently a number of new ionisation methods have been developed which now allow considerably more sequence data to be obtained from peptides. In particular, technical developments such as the use of fast atom bombardment (better known as FAB) and electrospray methods for producing volatile ions, have revolutionised the use of ms to study peptides and proteins. At present such technology is only available in specialist laboratories, but there is no doubt that mass spectrometry is providing a new generation of methods for studying peptides and proteins.

SAQ 4.2

You have purified a peptide. Amino acid analysis shows it to contain the following amino acids in the ratio shown: Asp_1, Glu_1, Lys_1, Arg_1, Phe_1, Ser_1, Ala_1

N-terminal analysis by the dansyl method showed the presence of aspartic acid. Treatment of the peptide with carboxypeptidase Y resulted in the release of serine. Cleavage of this peptide with trypsin (which cleaves on side of C-terminal arginine and lysine residues) gave three peptides (A, B and C). The amino acid composition and N-terminal amino acid for each of these three peptides, are tabulated below:

Peptide	N-terminus	Amino acid composition
A	Ala	Glu_1, Ala_1, Arg_1
B	Phe	Phe_1, Ser_1
C	Asp	Glu_1, Asp_1, Lys_1

Deduce the original sequence of the peptide. Start by trying to work out the sequences of peptides A, B and C.

SAQ 4.3

Which of the following statements is/are true?

1) Gramicidin A is a mammalian hormone.

2) Peptides can be fractionated using ammonium sulphate fractionation.

3) Peptides are smaller than proteins.

4) PTH amino acids can be analysed by reverse phase chromatography.

SAQ 4.4

You have isolated a peptide by reverse phase chromatography. The peptide eluted from the column as a single sharp peak, and analysis by treatment with dansyl chloride showed the presence of N-terminal valine only. You then decide to sequence the peptide using the Edman degradation and obtain the following results from your first four cycles.

Cycle number	PTH amino acid identified
1	valine
2	glycine and alanine
3	serine and threonine
4	leucine and valine

Can you explain what is happening? Why are you getting two PTH amino acids at cycles 2, 3 and 4?

Summary and objectives

The difference between a peptide and a protein is one of size; a short chain of less than 50 amino acids is known as a peptide; longer chains are called proteins. Peptides can be separated from proteins and nucleic acids (present in cells and tissue homogenates) by gel filtration. Reverse phase HPLC, using octadecyl silyl columns and exploiting the principle of ion pairing with trifluoroacetic acid (TFA), is used in peptide purification procedures. Elution is achieved with water/water-miscible organic solvent mixtures in which the concentration of organic solvent is gradually increased to remove the more hydrophobic peptides. The purity of a peptide is best assessed by the dansyl method for the determination of the N-terminal amino acid; The presence of a single N-terminal acid is indicative of a single pure peptide. The Edman degradation method, in which the N-terminal amino acids are removed sequentially, and the use of carboxypeptidase Y for cleaving C-terminal amino acids provide information about the amino acid sequence of a peptide. The limitations of these methods are discussed.

On completion of this chapter you should now be able to:

- describe the difference between a peptide and a protein;

- explain the principles of reverse phase HPLC as applied to the separation and purification of peptides;

- demonstrate the purity of a peptide using the dansyl method;

- describe the methods available for the sequencing of a peptide using the Edman degradation method for N-terminal amino acids and carboxypeptidase Y for C-terminal amino acids;

- use supplied data to determine the amino acid sequence of peptides.

Characterisation and structure determination of proteins

Characterisation and structure determination of proteins

5.1 Introduction

In the previous chapters we have considered the various ways in which proteins can be separated from biological materials and purified. We now have a purified protein and the next questions are firstly how do we characterise it and secondly what is its structure? These are topics which we will discuss in this chapter.

Characterisation of a protein is important in that is enables us to identify a particular molecule and compare it with samples from other laboratories. We shall also consider the determination of the partial sequence of amino acids in the molecule.

∏   Produce a list of the criteria you would use to confirm that you have a pure protein.

Normally a single band on a gel after electrophoresis (Section 1.3.4) and the presence of a single N-terminal amino acid (Section 4.2.3) would be sufficient evidence of purity. The inability to increase the specific activity of an enzyme is also a good indication of purity.

How much protein do you think you might end up with following a protein purification protocol:1 g, 100 mg, 1-10 mg or microgramme quantities?

Thirty years or so ago, when protein purification was in its infancy, not surprisingly workers concentrated on purifying those proteins that were present in great abundance, such as the blood protein haemoglobin and serum albumin, digestive enzymes such as chymotrypsin and trypsin, and the nuclear proteins, the histones. These workers were able to purify proteins with yields often in the range 100 mg -1 g.

Nowadays most of the readily available proteins have been purified, and protein chemists tend now to study proteins which are present at extremely low levels, such as nerve growth factor, tumour necrosis factor, interferons. Although some workers can purify a few milligrams of their protein, for others this is an undreamt of amount of protein; many are able to isolate only microgramme quantities of protein.

Fortunately, the recent development of highly sensitive methods for analysing proteins means that even with only microgramme amounts of protein, a considerable amount of information about the protein can be obtained.

Let us say you have purified 2 mg of your protein from the bacterium *Escherichia coli*. What are you going to do with it? Basically you must characterise your protein, ie identify properties, structural features, etc which allow you to differentiate this protein from all others, and then you will probably try and obtain some sequence data.

<table>
<tr><td>

SAQ 5.1

</td><td>

Professor Smith from America has written to you saying that he also has been studying *Escherichia coli* and he thinks he has isolated the same protein as you in his laboratory. What preliminary information would you require from him to allow you to compare his protein with yours? Try to list 4 or 5 different pieces of information.

</td></tr>
</table>

We will now consider each of these criteria.

5.2 Protein characterisation

biological
activity

Obviously there must have been a reason why Professor Smith thought you were studying the same protein and this is almost certainly because the biological activities of the two proteins are the same. There is little one can expand on here except to say, ensure that your assays are highly specific so that you can be sure that you and Professor Smith are both indeed studying the same activity. For example, you may both be working on an enzyme in which case you would expect your respective samples to catalyse the same reaction. Do ensure that the assay you are both using is totally specific for your enzyme. Remember that there may be more than one enzyme which either removes the substrate or produces the same final product (or both).

5.2.1 Relative molecular mass (RMM)

size

In an earlier chapter (Section 1.3.4) we saw a simple analytical method for separating proteins according to their size, ie relative molecular mass.

∏ Can you remember the name of this method?

gel
electrophoresis

It was SDS gel electrophoresis.

If at the same time as running your sample, you also run a number of other proteins of known RMM, by comparison of their relative mobilities one can deduce the RMM of your protein. You produce a calibration curve (which is essentially linear over the major portion of the graph) by plotting log RMM against distance moved. If you now measure the distance your protein moved through the gel you can read off its log RMM and hence you can calculate its RMM from the graph. The advantage of this method is that it uses little protein (~0.1 μg is needed to see a band on a gel) and you would be running a gel of your protein anyway to show it is pure, so why not include a few standard proteins on the same gel to give you an RMM value from the same experiment.

<table>
<tr><td>

SAQ 5.2

</td><td>

Table 5.1 shows the distance moved by a number of proteins of known RMM in a 10% polyacrylamide gel. On the same gel, a purified enzyme, aspartate aminotransferase (ATT) of unknown molecular mass, was also run. By plotting an appropriate graph determine the RMM of AAT.

</td></tr>
</table>

Protein	Relative molecular mass	Distance moved (mm)
transferrin	78 000	13
bovine serum albumin	66 000	21
ovalbumin	45 000	35
β-lactoglobulin	36 800	77
carbonic anhydrase	30 000	56
trypsinogen	24 000	64
myoglobin	17 800	80
cytochrome C	12 400	95
aspartate aminotransferase (ATT)	?	42

Table 5.1 Distance moved by a series of proteins in a 10% polyacrylamide gel.

In Chapter 1 a preparative method for separating proteins according to size was described.

∏ Can you remember the name of this method?

gel filtration

It was gel filtration. If you cannot remember this method refresh your memory by re-reading Section 1.3.4. In an approach analogous to the SDS gel method described above, by running a number of proteins of known RMM through a gel filtration column and measuring the elution volume of the proteins, a calibration curve can be constructed by plotting a graph of log RMM against elution volume. The elution volume of the protein of unknown RMM is also measured, and its RMM determined from the graph.

SAQ 5.3

Table 5.2 shows the elution volumes from a Sephadex G-100 column of a number of proteins of known RMM and the enzyme AAT. Plot a graph of log RMM against elution volume and determine the RMM of AAT. Compare the value you get with that obtained by SDS gel electrophoresis. Can you explain any differences between the two results?

Protein	Relative molecular mass	Elution volume (cm^3)
phosphorylase B	97 400	102
transferrin	78 000	121
bovine serum albumin	66 000	130
ovalbumin	45 000	154
β-lactoglobulin	36 800	170
α-chymotrypsin	22 500	197
lysozyme	14 300	235
aspartate aminotransferase (AAT)	?	118

Table 5.2 Elution volumes from a Sephadex-G-100 column of a number of proteins.

How good a calibration curve did you get from Table 5.1? Yes, it is quite a good calibration curve, and no, you were not given the wrong data for β-lactoglobulin - it really did move 77 cm in the gel. Can you explain why this point is so far off the graph (Figure 5.6)? To help you, use the calibration curve and the distance moved by β-lactoglobulin to determine the apparent RMM of this protein.

You should get an answer of about 18 400, in other words, about half the value given in Table 5.1. The reason for this is that β-lactoglobulin is another example of a protein that consists of two identical subunits. When run on an SDS gel we therefore see the running of the separate subunits and not the larger native protein comprising two linked subunits. We should therefore have used a RMM of 18 400 when constructing our calibration curve. However, in Table 5.2, 36 800 was the correct value to use since here we were looking at the behaviour of the native enzyme (ie the dimer).

Although many proteins are single polypeptide chains, a number of proteins consist of more than one polypeptide chain or subunit. We have already seen examples of this with β-lactoglobulin and aspartate aminotransferase. At least one other protein that we have mentioned so far in this chapter also exists as multiple subunits. Do you know which one it is?

It is haemoglobin. This protein comprises of four polypeptide chains: two identical α-chains and two identical β-chains.

Using SDS gel electrophoresis, Professor Smith obtained a value of 51 000 for the RMM of his protein. Using the same method, you obtained a value of 49 000.

∏ Does this mean the proteins are not the same?

These two values only differ by about 4%. This is well within the limits of accuracy of the method, and one would therefore be fairly confident that these two proteins, whether the same or not, have approximately the same RMM.

If you require a more precise value of the RMM of your protein then you must consider ultracentrifuge studies (Section 1.2.3). However, these require very expensive instruments and are not widely available.

5.2.2 Amino acid analysis

A full amino acid analysis of a protein involves determining the relative amount of each amino acid in the protein. To achieve this we must break all the peptide bonds in the protein to liberate the free amino acids, and then separate and measure the amount of each of these free amino acids. Can you remember how we do this? Refer to the basic procedure that was described earlier (Section 2.4) if you cannot remember.

ion exchange chromatography

To summarise briefly, the protein is broken down to free amino acids by hydrolysis in 6 mol l^{-1} hydrochloric acid. The free amino acids, in a known volume of hydrolysate, are then separated by ion exchange chromatography. As the individual amino acids elute from the bottom of the column, they are mixed with another reagent, traditionally ninhydrin, to give a blue coloured derivative the absorbance of which is measured by passing the column effluent stream through a spectrophotometer. Calibration curves of absorbance of known concentrations of amino acids are required to enable the amounts of the acids to be determined.

In recent years more sensitive detection reagents have been used such as o-pthalaldehyde and fluorescamine, both of which give fluorescent products with amino acids that can be detected and measured using a fluorometer. Methods that use ninhydrin, o-pthalaldehyde or fluorescamine are all referred to as post-column derivatisation methods, for fairly obvious reasons. An alternative approach is to derive the amino acids with a detectable 'tag' before loading onto the column. Such methods are called pre-column derivatisation methods, and include a reaction with o-pthalaldehyde to give a fluorescent product or the production of PTC-derivatives (formed by reaction with phenylisothiocyanate (PITC) which absorb at a wavelength of 280 nm and therefore their elution can be monitored at this wavelength).

∏ We have come across the use of PITC before. Can you remember what we used it for?

It was the reagent used in the first step of the Edman degradation for sequencing proteins and peptides (Section 4.3.1).

In post-column derivatisation methods, the amino acids are separated by ion-exchange chromatography (Section 1.3.4). However, in the pre-column derivatisation methods, the nature of the amino acid has been changed by derivatisation. The addition of o-pthalaldehyde and PITC both give considerable hydrophobic character to the amino acids. These derived amino acids are therefore separated by reverse phase chromatography (Sections 1.3.3 and 4.2.1) rather than ion-exchange chromatography.

reverse phase
chrom-
atography

Whichever of these methods is used, they are all sensitive down to at least the 0.1 nmol (100 pmol) level, and some are considerably more sensitive than this. In other words, you only need to sacrifice a maximum of 100 pmol of your protein to obtain an amino acid analysis. (Many workers quote sensitivity down to 10 pmol).

∏ Let us assume your protein has a RMM of 30 000. What mass of protein corresponds to 100 pmole?

1 mol has a mass of 30 000 g;

therefore 1 nmol has a mass of $30\,000 \times 10^{-9}\text{g} = 30\ \mu\text{g}$.

Therefore, 100 pmol has a mass of 3 μg.

We hope you are suitably impressed by the sensitivity of amino acid analysis. One can even cut stained protein bands out of gels (for example on SDS gel), hydrolyse the sample and carry out an amino acid analysis on this protein band.

Table 5.3 shows an amino acid analysis of your sample (protein A). Professor Smith's sample (protein B), and another sample (protein C) sent by Professor van den Brock.

Amino acid	Composition (mol %)		
	Protein A	Protein B	Protein C
Cys	1.8	1.7	2.0
Asx	7.4	7.3	9.8
Glx	5.3	5.2	7.2
Ser	8.1	8.4	3.4
Thr	4.2	4.0	5.9
Gly	10.4	10.6	14.7
Ala	5.8	6.0	11.8
Pro	4.9	4.7	4.0
Val	9.7	10.0	8.2
Met	0.8	0.9	1.0
Phe	6.2	6.4	5.1
Leu	8.1	7.9	4.9
Ileu	3.6	3.5	1.8
Tyr	6.4	6.1	3.2
His	0.6	0.5	4.2
Arg	7.3	7.1	5.8
Lys	9.4	9.7	7.0

Table 5.3 Amino acid analysis of proteins A, B and C. (Abbreviations of the amino acids are given in the appendices).

∏ From the information given in Table 5.3, do you think that either Professor Smith or Professor van den Brock have isolated the same protein as you?

You should be able to see from Table 5.3 that proteins A and B are nearly identical in their composition. This is very good evidence that the two proteins are the same. It is highly unlikely that two dissimilar proteins would give such similar amino acid analyses. Do no be worried by small apparent differences. For example, the values for valine seem to be different, being 9.7 and 10.0. However, this is only a difference of about 3% in the two figures which is well within experimental error. Also, either of the proteins, although essentially 'pure' may well contain traces of contaminating proteins which would make a small contribution to the overall amino acid analysis, and would contribute to a small difference such as this.

However, protein C is obviously different to proteins A and B. Just a casual look identifies major differences, for example in the composition of histidine, leucine and serine. There is no way that two identical proteins could give such dissimilar analyses, even allowing for a small amount of contaminating protein or a poor experimental procedure leading to considerable experimental error.

∏ Can you remember how many different amino acids are found in proteins? If not, look at Table 3.1 to remind yourself. How many different amino acids are shown in Table 5.3?

There are over twenty different amino acids found in proteins although only twenty are common. However, only seventeen are listed in Table 5.3.

⊓⊓ Why is there this difference? Compare the list of amino acids in Table 3.1 with those shown in Table 5.3 and write down any differences that you notice.

tryptophan

Firstly, you will have noticed that tryptophan (Trp) is missing from Table 5.3. This is because tryptophan is totally destroyed at the acid hydrolysis step used to produce the mixture of amino acids (commonly referred to as the protein hydrolysate). Since tryptophan is a relatively uncommon amino acid this is not a great loss to our overall analysis. However, the tryptophan content can be estimated in the original protein by taking certain measurements. This is based on a characteristic of tryptophan that is important in protein chemistry. Can you remember what is it? If not, refer to Section 2.2.1 to refresh your memory.

Tryptophan has a characteristic UV spectrum with an absorption maximum at 280 nm. (Remember that this is one of the ways that we monitor proteins in solution). By applying appropriate equations one can determine the tryptophan content from the A_{280} of the native protein.

asparagine
glutamine

Secondly, you will have noticed that neither asparagine (Asn) and aspartic acid (Asp), nor glutamine (Gln) and glutamic acid (Glu) appear in the list, whereas Asx and Glx, which are not shown in Table 3.1, do appear. What are Asx and Glx? Again, they arise as a consequence of the acid hydrolysis step.

⊓⊓ Look at the side chains of asparagine and glutamine in Table 3.1. What do you think might happen to these at the acid hydrolysis step?

They are deaminated to the corresponding carboxylic acid. This is shown below for asparagine, which is deaminated to aspartic acid.

$$
\begin{array}{ccc}
\begin{array}{c}
CONH_2 \\
| \\
CH_2 \\
| \\
H_2N - C - COOH \\
| \\
H
\end{array}
&
\xrightarrow{\ H^+\ }
&
\begin{array}{c}
COOH \\
| \\
CH_2 \\
| \\
H_2N - C - COOH \quad + \quad NH_3 \\
| \\
H
\end{array}
\\
\text{asparagine (Asn)} & & \text{aspartic acid (Asp)}
\end{array}
$$

Following acid hydrolysis therefore, neither asparagine nor glutamine are found in the hydrolysate: they have been converted to aspartic and glutamic acids respectively. The amino acid analysis therefore detects only aspartic and glutamic acid. Since it is not possible to determine how much of each of these originally existed as the acid or the amide, we present the data as Asx and Glx.

For example, in protein A (Table 5.3), 7.4% of all the amino acids are a mixture of asparagine and aspartic acid, but we cannot tell what the relative amounts of each are. At one extreme, 7.4% of the amino acids may be asparagine, with no aspartic acid present in the protein. At the other extreme, 7.4% of the amino acids may be aspartic acid, with no asparagine in the protein.

However, the truth will probably be somewhere in the middle since proteins tend to contain both asparagine and aspartic acid, and similarly both glutamine and glutamic acid. Only when the amino acid sequence is determined for the protein will we know the exact number of acids and amides.

Both of the above problems arose from the acid hydrolysis step. Further minor problems also exist. For example, about 10% of the serine is destroyed at this step. On the other hand, peptide bonds between hydrophobic amino acids (eg Leu-Ileu) are incompletely cleaved, leading to these amino acids being underestimated by a few percent. However, this should not be a problem if we are comparing two analyses, as we have been doing above, since these small errors will be the same for both samples, which should therefore still give the same overall amino acid analyses.

5.2.3 Peptide mapping

protein cleaving

Peptide mapping involves cleaving the protein (usually using an enzyme) to give a number of peptides which are then separated to give a 'map' or fingerprint of that protein. Obviously if we carry out the same procedure on two proteins that we think are the same, then if they are the same they will give identical maps. The first step is to cleave the protein into peptides.

The enzyme trypsin is commonly used as this cleaves the peptide bond after every lysine and arginine residue in a protein, but nowhere else. A series of discrete peptides are therefore formed, each one ending (ie C-terminal group) with arginine or lysine.

∏ Can you remember what type of amino acids lysine and arginine are (eg neutral, hydrophobic, charged)?

They are both basic amino acids. In other words they have positively-charged side chains.

Alternatively, endoproteinase Glu-C from *Staphylococcus aureus* can be used. This cleaves after every aspartic and glutamic acid residue.

∏ What type of amino acids are aspartic and glutamic acids?

The name is a bit of a giveaway! They are both acidic amino acids. They each have a negatively-charged carboxyl group on their side chain. Note that endoproteinase Glu-C is called endopeptidase Glu-C by some authorities.

Approximately how many peptides do you think we might produce from a given protein by cleavage with an enzyme?

Consider a purified protein of RMM 50 000, since the average RMM of an amino acid is 110, the number of peptide residues obtained will be given by 50 000/110, ie approximately 450.

∏ The amino acid analysis for a protein (protein A) is given in Table 5.3, using this information together with the above calculations work out how many peptides you should get from protein A following cleavage with trypsin.

Table 5.3 shows you that 16.7% of all the amino acids in the protein are either arginine or lysine. In other words for every 100 amino acids, 16.7 are arginine or lysine.

Since there are 450 amino acids, 4.5 x 16.7 = 75 amino acids are arginine or lysine. Since trypsin should cleave after each of these; this should generate 76 peptides. Obviously this calculation contains some approximations, particularly with respect to the number of amino acids in the protein, but we can certainly say that there should be about 75 tryptic peptides produced.

peptide
separation

How do we go about 'spreading out' these peptides to give a peptide map? Traditionally this has been done in two dimensions on a cellulose thin-layer chromatography plate, usually of dimensions 20 x 20 cm. The protein digest is loaded in one corner of the plate (Figure 5.1), the plate lightly sprayed with buffer solution to wet it and then the sample subjected to electrophoresis along one edge of the cellulose plate (Section 1.4.1). The peptides are therefore separated according to their different charges. The plate is then dried in a stream of warm air, placed in a chromatography tank and chromatographed at right angles to the direction of electrophoresis. Following chromatography, the plate is dried and the peptides detected.

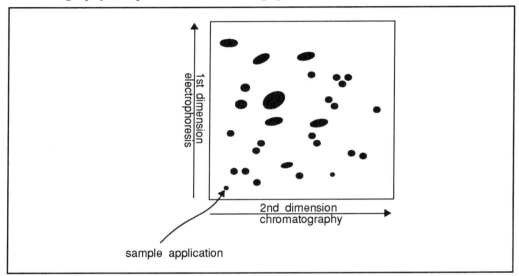

Figure 5.1 Peptide mapping on a 2D TLC plate.

∏ Can you remember what methods we can use to detect peptides on a TLC plate?

The traditional method is spraying with ninhydrin followed by heating at 100°C for 10 minutes Peptides are identified as blue spots. A more sensitive method is to spray with fluorescamine which reacts with N-terminal amino acids to give fluorescent derivatives which can be viewed under UV radiation.

We said above that the 2D mapping on a chromatography plate is the traditional method, and although still a very good method, it has generally been superseded by a one-dimensional method.

∏ Can you think of a sensitive and accurate method for separating peptides? If you cannot, refresh your memory by reading Section 4.2.1.

The method is reverse phase chromatography. The mixture of peptides could certainly be resolved into a minimum of 40-50 peaks which would be a totally acceptable 'fingerprint' of your protein. This method also has the advantages of being quick (approximately 30 minutes) compared to 2D mapping (one working day) and requires considerably less material., 1-2 μg against 200-500 μg for the 2D method.

Peptide mapping has been described as a method for comparing two purified proteins to see if they are the same. This is certainly one of the uses of peptide mapping. There are, however, others which we will discuss in subsequent sections.

5.2.4 Applications of peptide mapping

sickle-cell anaemia

Peptide mapping is also used to identify minor differences in the analogous protein from different sources. The classic example of this was the analysis of haemoglobin from patients suffering from sickle-cell anaemia. The symptoms of this disease, including the 'sickle' shape of the patients' red blood cells (and hence the name of the illness), are due to the fact that these patients have a modified form of haemoglobin, Haemoglobin S or HbS. The nature of this change was determined by Vernon Ingram in 1954 using two-dimensional peptide mapping on a sheet of filter paper.

The haemoglobin molecule comprises two identical α-chains and two identical β-chains. The α- and β-chains have a total of 27 arginine and lysine residues. Ingram cleaved these with trypsin to produce 28 peptides, which were then separated on a sheet of filter paper by electrophoresis in the horizontal direction and chromatography in the vertical direction, and identified by staining with ninhydrin. When he compared the peptide maps from normal haemoglobin (HbA) and HbS he found that all but one of the peptide spots matched. The one spot that was different was eluted from each map and each shown to be a single peptide of 8 amino acids. When the peptides were sequenced the results showed that HbS contained valine instead of glutamic acid at position 6 of the β-chain:

HbA: Val-His-Leu-Thr-Pro-Glu-Glu-Lys-;

HbS: Val-His-Leu-Thr-Pro-Val-Glu-Lys-.

∏ Would you say that the replacement of glutamic acid by valine is a conservative change or has the nature of the amino acid at this position been changed dramatically? (A conservative change is one where the physical character of a side chain is not changed, ie hydrophobic residue replaced by another hydrophobic residue, eg leucine replacing isoleucine).

The replacement of a charged amino acid by a hydrophobic amino acid is certainly not a conservative change (it would have been if glutamic acid had been replaced by aspartic acid). The nature of the amino acid at this position has been changed dramatically.

This simple replacement of a single acid amino acid by a hydrophobic amino acid may seem like a very small change to the overall structure of haemoglobin, but it considerably affects the function of the haemoglobin molecule which in turn has dramatic effects on the person carrying this mutated gene.

The symptoms for those carrying the gene for HbS include anaemia, shock, cardiac enlargement, jaundice and kidney damage.

SAQ 5.4

What would you expect to see if you compared samples of haemoglobin A and haemoglobin S on an SDS gel? How many bands would you get for each sample? Would you be able to detect the differences between HbA and HbS?

5.2.5 Cleveland peptide mapping

Cleveland peptide mapping

One further form of peptide mapping has been introduced in recent years, known as Cleveland peptide mapping, named after the originator of the method. Let us assume you have purified and studied a protein from bovine tissue. You now wish to investigate whether this protein exists in tissue from other species, say chicken. You therefore run an SDS gel of your purified protein against an extract of chicken tissue. This is shown in Figure 5.2. As you can see, the chicken tissue extract contains a protein band that runs in the same place as your purified protein.

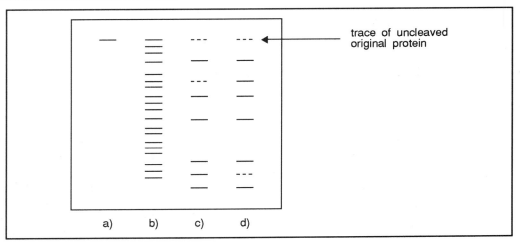

Figure 5.2 Cleveland peptide mapping. a) Your purified protein; b) chicken tissue extract; c) and d) Cleveland peptide maps of two separate protein bands cut out of a gel. The great similarity in the peptide patterns suggests that the two proteins are the same. (See text for a discription of the Cleveland technique).

Would this convince you that your protein is present in chicken tissue? If there was not a band in this position we could be fairly certain that your protein was not present in chicken tissue. However, the band that you have seen does not necessarily have to be the same protein - all we can say is that the chicken extract contains a protein of the same RMM as yours. It may well be that it is the same protein, but you have not proved it.

One way forward would be to go ahead and purify this protein, but what a waste of time it would be if after six months or a year you have the purified protein and find it is not the one you want! You would feel far more confident working on the purification of this protein if you had much better evidence that this was the chicken analogue of your protein.

Cleveland mapping can give you this evidence.

Both of the stained bands that you wish to compare are cut out of the gel, equilibrated in an appropriate buffer solution and placed in adjacent wells of another SDS gel. The gel pieces are then overlayered with a solution of protease (normally chymotrypsin which retains its activity in the presence of SDS). A potential gradient is then applied whereupon the protein moves out of the stained band into the stacking gel where it is concentrated into a sharp band together with the protease. At this stage the power is turned off for about 20 minutes to allow the protease to digest the protein. The amount of protease used is very low to ensure that the protein is only cleaved in a few places, giving rise to large peptides. The potential is then applied again and electrophoresis continued as usual. As the peptides move through the separating gel they separate according to size, so that when the gel is stained we see a number of bands in each track. If the two proteins cut out of the original gel are the same, they should show very similar peptide patterns. If they were not the same, the patterns would be quite different.

∏ Supposing too much protease was added in this method, what would happen?

In this case the proteins would be digested down to very small peptides of RMM of, say, 1000-2000. You may remember that even a 15% polyacrylamide gel only separates proteins down to RMM of about 10 000. Proteins or peptides smaller than this travel at the electrophoresis front as a single band, and often wash out of the gel during the staining and destaining steps. The use of too much protease would therefore result in no bands being seen on the gel.

5.3 Protein sequence determination

The methods we have discussed above are all relatively quick methods (none taking more that one day) for providing useful information which allows you to characterise your protein and compare it with others. However, the ultimate characterisation of a protein is the determination of its complete amino acid sequence, the primary structure of the protein, and ultimately its complete three-dimensional (tertiary) structure.

primary structure

The determination of a complete amino acid sequence is not a task to be undertaken lightly. Depending on the facilities available, it is not unusual for it to take one to two years' work to determine the primary structure of a protein, and sometimes longer. Also, it usually requires milligramme quantities of protein to achieve this task. However, such information is essential to help determine the three-dimensional structure of the protein and hence, hopefully, understand the relationship between its structure and function.

However, a complete amino acid sequence is not always necessary. For many people, determination of part of the sequence of a protein will provide sufficient information for the task at hand. For example if you can obtain an N-terminal sequence of your protein of say 20 residues and can do the same for a different sample, this data alone should tell you if the two proteins are the same. If they are, the sequences may also be sufficient to provide enough information to construct a probe to isolate the gene for the protein. The gene may then be sequenced and the amino acid sequence of the protein deduced from the gene sequence. For those who can only isolate trace amounts of their protein, and thus obtain only limited amounts of protein sequence data, this approach of gene isolation and sequencing is a much more viable one than trying to obtain the total sequence by direct sequencing of the protein.

Π   There is a chemical method for removing amino acids one at a time from the N-terminus of a polypeptide. Can you remember what the method was called?

Edman degradation

It was the Edman degradation (Section 4.3.1).

It would appear therefore that sequencing a protein should be an easy task. One simply starts sequencing at the N-terminus using the Edman degradation and continues for a few hundred cycles until one reaches the end of the protein.

Unfortunately it is not that easy. The Edman degradation is not 100% efficient and leads to the problem of 'staggered ends'. Under optimal conditions (see below) a repetitive yield of 98-99% can be achieved, but even this limits the amount of useful sequence data from a single run to about 60 or so residues.

5.3.1 Enzymic cleavage of proteins

It should be obvious from this that we cannot sequence our protein of 450 residues in one go. The approach adopted therefore is to breakdown the protein with an enzyme, for example, trypsin, into a number of smaller peptides (ideally 5-15 residues long), which can then be purified and individually sequenced manually (in a test-tube) using the Edman method.

For example, we have seen that for our hypothetical protein (protein A, Table 5.3) we would have about 75 tryptic peptides. Even if we purify and sequence all these peptides, we would still not know in what order to join these peptides. One therefore cleaves some more protein, using an enzyme that cleaves in a different position, eg endoproteinase Glu-C, to produce a different set of peptides. These would also be purified and sequenced, and this should provide overlapping sequences which allow some of the tryptic peptides to be joined together. It may be necessary to carry out four or five different digests of the protein to provide sufficient overlaps to obtain the complete sequence. Some of the enzymes more commonly used to produce peptides are shown in Table 5.4.

Enzyme (source)	Cleavage site
chymotrypsin (cow pancreas)	C-terminal to hydrophobic residues, eg Phe, Tyr, Trp, Leu
clostripain (the micro-organism *Clostridium histolyticum*)	C-terminal to Arg residues
elastase (pig pancreas)	C-terminal to amino acids with small hydrophobic sidechains
endoproteinase Arg-C (mouse submaxillary gland)	C-terminal to Arg residues
endoproteinase Glu-C (the micro-organism *Staphylococcus aureus*)	C-terminal to Asp and Glu
endoproteinase Lys-C (the micro-organism *Lysobacter enzymogenes*)	C-terminal to Lys
pepsin (pig stomach)	broad specificity but preference for cleavage C-terminal to Phe, Leu and Glu
thermolysin (the micro-organism *Bacillus thermoproteolyticus*)	N-terminal to amino acids with bulky hydrophobic sidechains, eg Ileu, Leu, Val and Phe
trypsin (cow pancreas)	C-terminal to Lys and Arg

Table 5.4 Some commonly used proteolytic enzymes.

The following SAQ should clarify the principle of the use of overlapping peptides for you.

SAQ 5.5

The following four peptides were isolated from a tryptic digest of a protein.

1) Ser-Ala-Val-Leu-Gly-Asp-His-Phe-Arg;

2) Gly-Val-Glu-Glu-Met-Trp-Thr-Lys;

3) Ala-Pro-Ser-Glu-Val-Thr-Gly-Phe-Leu-Ileu-Arg;

4) Val-Ala-Gly-Gly-Leu-Thr-Cys-Ala-Lys.

The peptide His-Phe-Arg-Ala-Pro-Ser-Glu was isolated from an endoproteinase Glu-C digest of the same protein. Try and show how this peptide data can be used to construct a larger length of sequence from the protein (not the complete protein).

This therefore is how proteins have been traditionally sequenced. It is necessary to generate small peptides because the Edman degradation (Section 4.3.1) when carried out in a test tube, is only 93-95% efficient under these conditions and therefore 15-20 residues is the very maximum useful sequence data one can get from a peptide. Obviously such an approach is very time consuming and requires quite a lot of protein.

10-20 years ago it was not unusual for it to take 5-10 man years to sequence a protein, requiring hundreds of milligrams of protein. With recent improvements in methods for isolating peptides, especially the introduction of reverse phase HPLC, the task can now be achieved in 1-3 man years, often requiring only tens of milligrams of protein.

Figure 5.3 shows the complete amino acid sequence of sperm whale myoglobin.

a)

```
                                                                        20
     H-ValLeuSerGluGlyGluTrpGlnLeuValLeuHisValTrpAlaLysValGluAlaAspVal
                                                                          Ala
                    40                                                    Gly
  ArgAspPheLysGluLeuThrGluProHisSerLysPheLeuArgIleLeuIleAspGlnGlyHis
  Phe
  Lys                                       60
  HisLeuLysThrGluAlaGluMetLysAlaSerGluAspLeuLysLysHisGlyValThrValLeu
                                                                          Thr
                                                 80                       Ala
 AlaHisSerGlnAlaLeuProLysLeuGluAlaGluHisHisGlyLysLysLysLeuIleAlaGlyLeu
 Thr
 Lys       100                                                         120
 HisLysIleProIleLysTyrLeuGluPheIleSerGluAlaIleHisValLeuHisSerArgHisPro
                                                                         Gly
          140                                                            Asn
AlaAlaIleAspLysArgPheLeuGluLeuAlaLysAsnMetAlaGlyGlnAlaAspAlaGlyPhe
Lys
Tyr
LysGluHO-LeuGlyTyrGlnGly-OH
```

b)

```
                             20
   H--V-L-S-E-G-E-W-Q-L-V-L-H-V-W-A-K-V-E-A-D-V
                                            A
                 40                         G
     R-D-F-K-E-L-T-E-P-H-S-K-F-L-R-I-L-I-D-Q-G-H
     F
     K                      60
     H-L-K-T-E-A-E-M-K-A-S-E-D-L-K-K-H-G-V-T-V-L
                                              T
                             80               A
   A-H-S-Q-A-L-P-K-L-E-A-E-H-H-G-K-K-K-L-I-A-G-L
   T
   K   100                                  120
   H-K-I-P-I-K-Y-L-E-F-I-S-E-A-I-I-H-V-L-H-S-R-H-P
                                              G
        140                                   N
  A-A-I-D-K-R-F-L-E-L-A-K-N-M-A-G-Q-A-D-A-G-F
  K
  Y
  K-E(HO)-L-G-Y-Q-G---OH
```

Figure 5.3 Amino acid sequence of sperm whale myoglobin: a) with normal abbreviations and b) with single letter code (see appendices for abbreviations and nomenclature of amino acids).

Fortunately, developments in the automation of the Edman degradation in recent years have considerably speeded up the sequencing process, and perhaps more importantly have greatly increased the sensitivity of the method such that sequence data can now be obtained from extremely small amounts of protein, often as little as one microgram of protein.

protein sequencer

The automation of the Edman degradation resulted in the development of the 'protein sequencer'. The current state-of-the-art protein sequencer is the 'gas-phase' protein sequencer so named because one of the reagents in the Edman degradation are introduced to the protein in the gas phase. Such a machine carries out one cycle of the Edman degradation in about 30 minutes, automatically converts the thiazolinone to the more stable PTH derivative and identifies the derivative by injecting it onto an HPLC column. The machine is highly sensitive, being able to obtain sequence data from as little as 10-100 pmol of protein. Because of the design of the machine, the repetitive yield can be as high as 98-99%, so sequences of sixty or more residues can be elucidated in a single run under favourable circumstances. It should be stressed that such machines are highly expensive and tend to only be available in specialist laboratories.

5.3.2 Chemical methods of cleavage of proteins

With the advent of the protein sequencer, our strategy has therefore changed. We now wish to produce a small number of larger peptides for automatic sequencing rather than the large number of small peptides for manual Edman sequencing. Enzymes are not very useful for producing large peptides, since there are often many cleavage sites in a protein. However for two of the least abundant amino acids, methionine and tryptophan, there are chemical methods for cleaving at these residues, resulting in large peptides.

chemical methods for cleaving

For example, treatment of proteins with cyanogen bromide causes cleavage at methionine residues.

Π For our hypothetical protein A of 450 residues, from Table 5.3 calculate the number of peptides we would expect from a cyanogen bromide cleavage of this protein.

There are 0.9 methionine residues per 100 amino acids. Therefore in 450 amino acids we would expect to find 4.5 x 0.9 = 4.05 methionine residues. With four methionine residues, cleavage at each residue would result in five peptides, average length of 90 residues each. These peptides would be ideal for automated sequence analysis. Also, it should not prove to be a difficult task to purify a mixture of only 5 peptides (or perhaps we should call it a mixture of five small proteins).

Using modern day technology, it is possible therefore, even with only microgram quantities of protein, to obtain an N-terminal sequence of 30-60 residues, and often to also produce large peptides which allows one to determine further the sequence of internal regions of, say, 20-40 residues at a time. A reasonable part of the total sequence of a protein can therefore be built up in a relatively short time using small amounts of protein.

With the ability to obtain some sequence data from only a trace amount of a protein, even a band extracted from an SDS gel should generate sufficient sequence data, (even if only part of the N-terminal sequence) to construct a DNA probe which should enable us to isolate the gene for the protein. This would then allow us to sequence the gene, thus obtaining the full protein sequence, and also to clone and express the gene, thus producing large quantities of the protein for further studies.

5.4 The three-dimensional structure of proteins

A detailed discussion of the three-dimensional structure of proteins is beyond the scope of this book and the reader is referred to the BIOTOL text, 'The Molecular Fabric of Cells,' for a full description.

We remind you, however, that we can divide the structure of proteins into three main orders (illustrated diagrammatically in Figure 5.4).

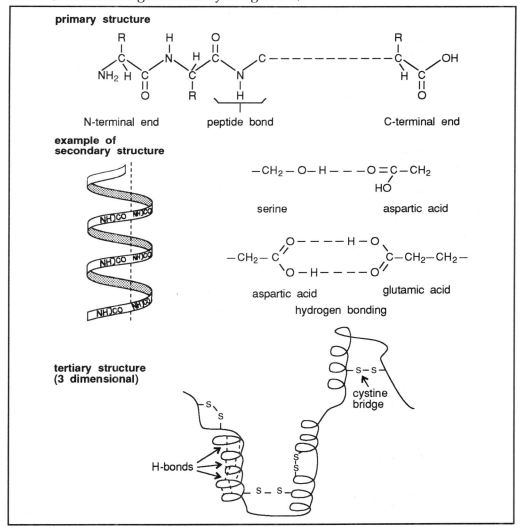

Figure 5.4 Primary, secondary and tertiary structures of proteins. A protein chain is composed of amino acids joined together by peptide bonds, and has a free amino group at one end (the N terminal end) and a free carboxyl group at the other (the C terminal end). The chains can form themselves into a helical secondary structure, each turn of the helix involving about four amino acids, and the helix is kept in place by hydrogen bonds between the amino acids of the adjacent turns. Note that other arrangements of peptide chains do exist (eg, β pleated sheets) in some proteins. (See the BIOTOL text, 'The Molecular Fabric of Cells,' for further details). The tertiary structure describes the way in which the areas of secondary structure are folded and held together to give a defined three-dimensional shape.

primary structure

- the primary structure is the actual sequence of the amino acids within the peptide chains;

secondary structure

- the secondary structure is the arrangement of the peptide chains into regular arrays (eg α-helixes, β-pleated sheets);

tertiary structure

- the tertiary structure describes the way in which the areas of the secondary structure are folded and held together in definite shapes in three-dimensional space.

The molecular interaction and the physical forces which cause proteins to take up particular conformations are illustrated in Figure 5.5.

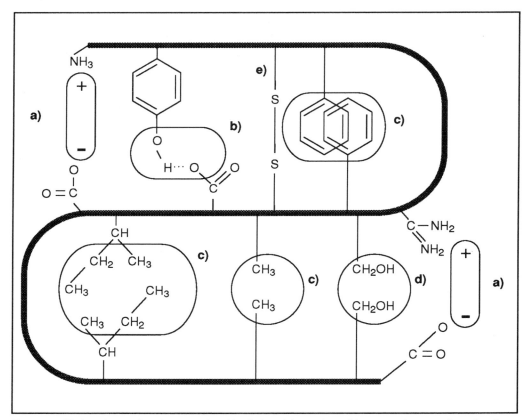

Figure 5.5 Various interactions between the side chains of the amino acids present in a protein molecule. The interactions are: a) electrostatic force; b) hydrogen bonding; c) hydrophobic interactions; d) van der Waals interactions and hydrogen bonding; e) disulphide bridge.

In this brief section we will focus on the methods for the determination of these three-dimensional structures.

To date, 3D structures of protein have been determined by X-ray crystallography. Crystallisation of a protein is often said to be more of an art than a science, but can be achieved, and gene cloning and the expression of these genes nowadays will give us sufficient of any protein to contemplate attempting to produce crystals. The complex diffraction patterns obtained when X-rays are passed through a protein crystal require extensive and detailed mathematical analysis but modern computer technology has gone a long way to speeding up this process.

A more recent development has been the determination of protein structure, in solution, by nuclear magnetic resonance spectroscopy (NMR). Currently NMR can be used to study proteins up to a RMM of 20 000. The advantage of this method is that the protein is studied free in solution (unlike the more restrictive environment in the crystal) and can also analyse conformational changes due to the binding of ligands, substrates, drugs, etc.

Both NMR and crystallography rely heavily on the need for protein sequence data to help interpret results. The production of protein sequence data is, therefore, an important role for the protein chemist, be it to obtain a complete primary structure, for use in the interpretation of NMR or crystallography data, or a short length of sequence sufficient to construct a probe for gene isolation, sequencing, cloning and expression.

SAQ 5.6

You have purified the polypeptide shown below, and shown the polypeptide to have two quite different biological activities. You have reason to believe that the N-terminal region is responsible for one of these activities and the C-terminal region for the other activity.

```
            5                10               14
            |                |                |
Ala-Gly-Arg-Ser-Ala-Glu-Phe-Lys-Gly-Pro-Glu-Lys-Pro-Ala
```

Which of the following enzymes would you use to cleave the peptide in the central region so that you could separate and purify the N-terminal region and the C-terminal region and thus study them separately? Explain the reason for your choice. You may need to refer to Table 5.4 to refresh your memory on the cleavage sites of these enzymes: a) trypsin; b) chymotrypsin; c) endoproteinase Glu-C.

Summary and objectives

A purified protein can be characterised by a range of methods, such as biological assay, RMM determination, amino acid analysis and peptide mapping. Amino acid sequencing is essential to the determination of the structure of a peptide and hence a protein. Protein sequencing is achieved by first cleaving the protein by enzymatic or specific chemical methods followed by peptide mapping. X-ray crystallographic and NMR spectroscopic methods are used to determine the 3D structure of a protein in the crystalline form and in solution respectively.

On completion of this chapter you should be able to:

- determine the relative molecular mass of a protein using data obtained from SDS-gel electrophoresis and Sephadex gel filtration;

- explain the principles of amino acid analysis and sequencing and peptide mapping;

- describe the ways in which a protein can be cleaved by enzymes and specific chemicals;

- discuss the strategy employed in sequencing a protein;

- use supplied data to amino acid sequences in peptides;

- give a brief account of the methods available for the determination of the 3D structure of a protein.

Introduction to the study of nucleic acid

Introduction to the study of nucleic acid

6.1 Introduction

The importance of nucleic acids in biological systems is well established. They play essential roles in the storage of genetic information (DNA) and in the conversion of this genetic information into the amino acid sequences found in proteins. The purpose of this chapter is not to review these roles but to remind you of the important chemical and physical features of these molecules and to introduce you to some important techniques which underpin their study. In subsequent chapters we will build upon this knowledge by examining more specifically the techniques that have been applied to the extraction, estimation, purification and analysis of these important molecules. A more detailed discussion of the structure and occurrence of nucleic acids is covered in the BIOTOL text, 'The Molecular Fabric of Cells,' and their functions are described in the BIOTOL text, 'The Infrastructure and Activities of Cells.' Here, the main theme is to examine the technique for studying these molecules.

6.2 Chemical composition of the nucleic acids

The nucleic acids, polymers of repeating units, are compounds of high molecular mass found in association with proteins, the nucleoproteins. The two main groups of nucleic acids are the ribonucleic acids (RNA) and the deoxyribonucleic acids (DNA). Hydrolysis of nucleic acids gives, in the first instance, nucleotides; these are the repeating units of the nucleic acids. Further hydrolysis gives nucleosides and phosphoric acid and eventually a pentose sugar and a base; this is summarised in Figure 6.1.

nucleotides
nucleosides
pentose

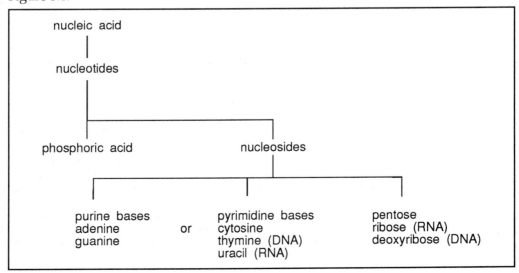

Figure 6.1 The relationship between nucleic acid and its low molecular mass components.

We will consider first the structure of the compounds of lowest molecular mass, the bases and pentoses, and then see how these are combined to form nucleosides, nucleotides and eventually the nucleic acids. purines

Purine and pyrimidine bases are shown in Figure 6.2. Adenine (A) and guanine (G) are substituted purines; these are present in all nucleic acids.

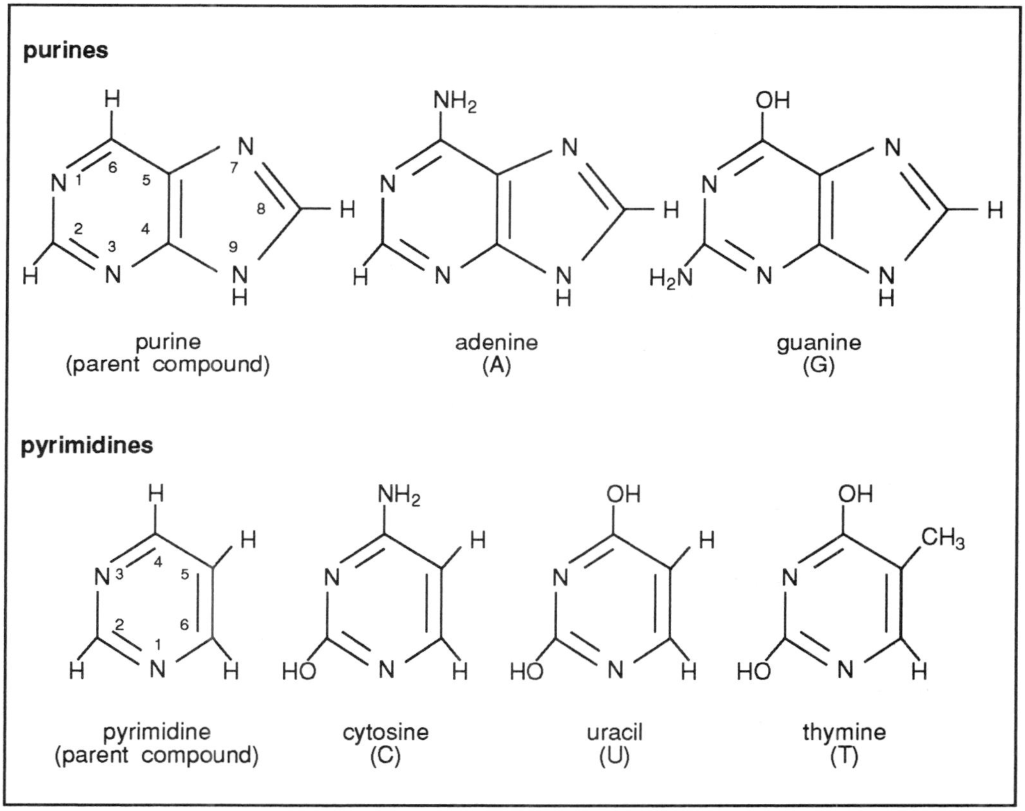

Figure 6.2 Purine and pyrimidine bases. Note that the hydrogens of the hydroxyl groups can undergo tautomeric shift to form keto (oxo) groups, (see for example the illustration of uracil in Figure 6.4).

pyrimidines

Cytosine (C), found in both RNA and DNA, uracil (U), found only in RNA, and thymine (T), found only in DNA are substituted pyrimidines.

pairing of complementary bases

Pairing of complementary bases (purine-pyrimidine) by hydrogen bonding (see Figure 6.8) results in the formation of the double helix structure in DNA (see Figure 6.9). All normal DNA molecules exhibit certain chemical regularities: adenine always pairs with thymine and guanine with cytosine. Thus as a corollary it follows that the sum of all the purine bases $\Sigma(A + G)$ is equal to the sum of all the pyrimidine bases $\Sigma(C + T)$.

pentose sugars

Pentose sugars are shown in Figure 6.3. The two main types of nucleic acids derive their names from the sugars present, which is either ribose or deoxyribose. The carbon atoms on the sugars are denoted as 1', 2' etc to differentiate them from the atoms of the bases.

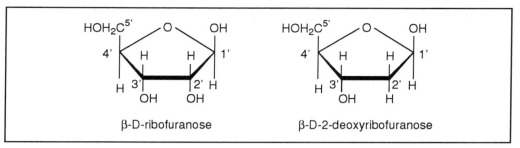

Figure 6.3 The pentose sugars.

nucleosides   Nucleosides are formed when the C1 of the sugar is linked to the nitrogen in the 9 position for purines or the 1 position for pyrimidines. Most of the bonds linking the sugar and the base are as shown in Figure 6.4 but transfer RNA (tRNA) contains an unusual nucleotide in which the ribose is linked to the uracil via the 5' position.

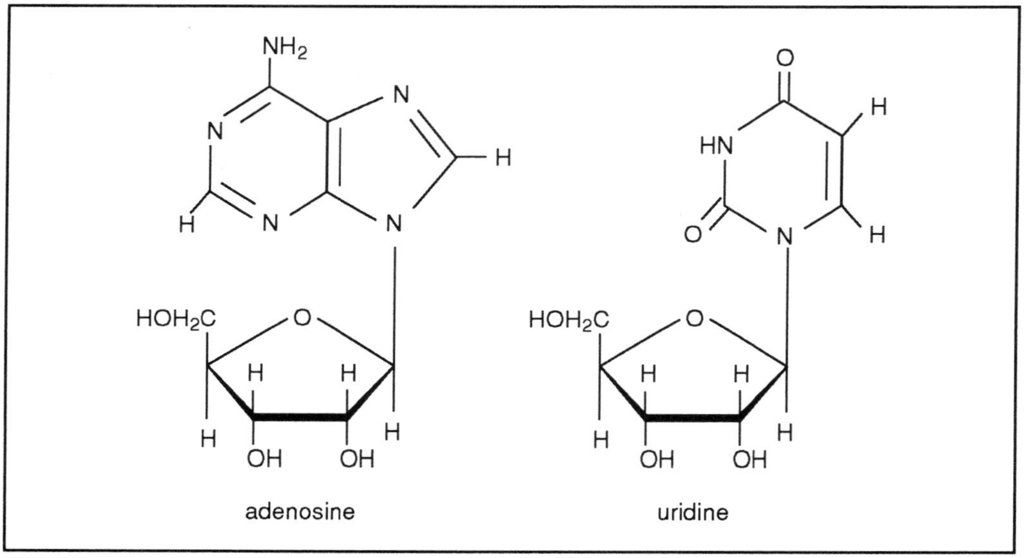

Figure 6.4 Typical nucleosides.

nucleotides   Nucleotides are sugar-phosphate esters of the nucleosides (Figure 6.5); they are strong dibasic acids.

The hydroxyl groups in positions 2', 3' and 5' of ribose and 3' and 5' or deoxyribose can be esterified with phosphoric acid; all these esters are known. Nucleotides and nucleosides are named after the bases contained in their structure, thus for adenine the nucleoside is adenosine and the nucleotide is 5'-adenylic acid.

Other biologically important nucleotides which occur free in nature include adenosine di- and tri-phosphates (ADP, ATP) and nicotinamide adenine dinucleotide (NAD). These are discussed elsewhere in the BIOTOL series of texts ('Principles of Cell Energetics').

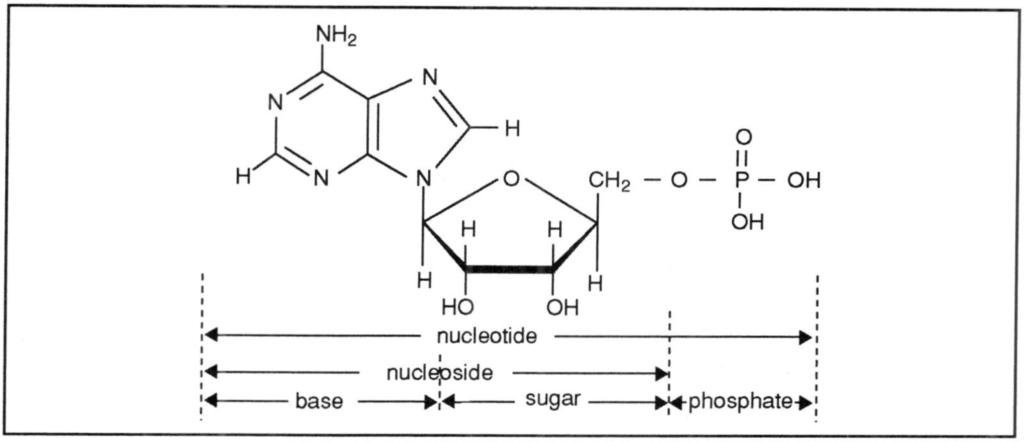

Figure 6.5 A typical nucleotide: 5' adenylic acid.

Nucleic acids Nucleic acids are polynucleotides in which the nucleotides are linked by phosphodiester bonds between the 3' and 5' position of the sugar moieties. A portion of a molecule of RNA therefore has the structure given in Figure 6.6 where the base is either a purine or a pyrimidine.

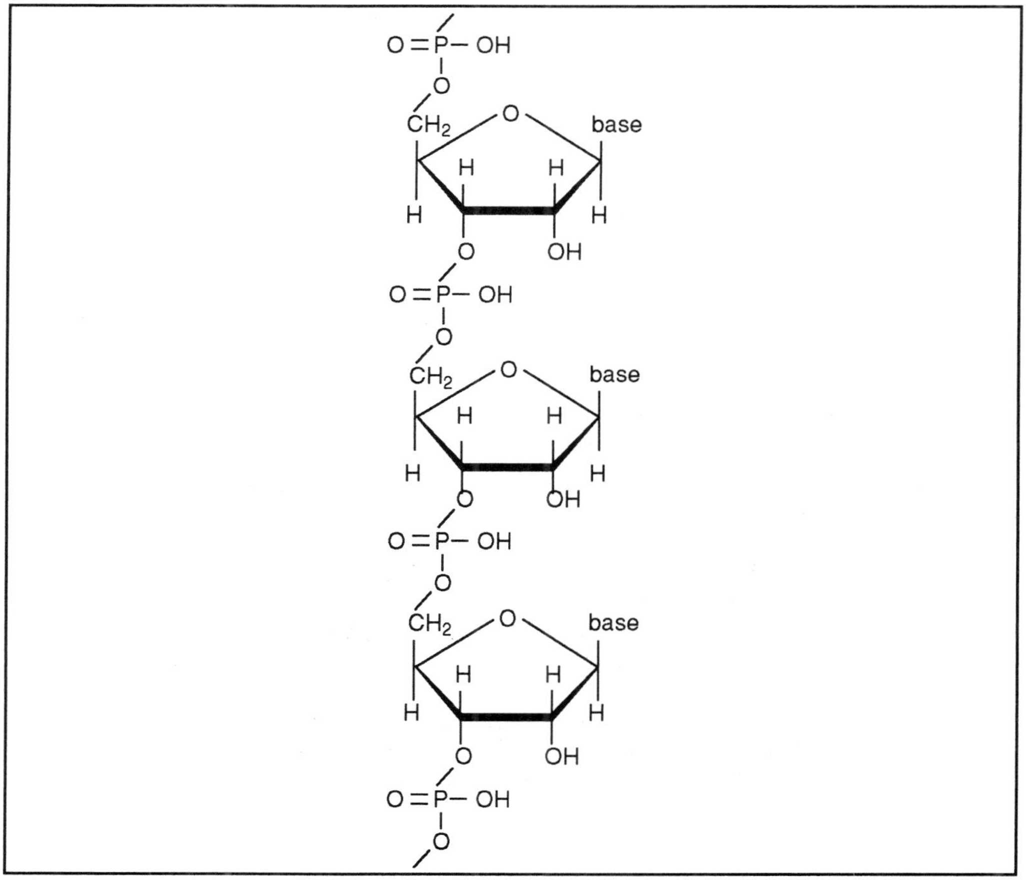

Figure 6.6 Part of the molecule of RNA.

A useful form of shorthand for the structure showing the bases present is given in (Figure 6.7). By convention, the sequence of bases in a RNA molecule is written 5' to 3' using the single letter codes A, G, C and U. This is because of the way in which the information carried by RNA molecules is read (for more details see BIOTOL text, 'The Infrastructure and Activities of Cells').

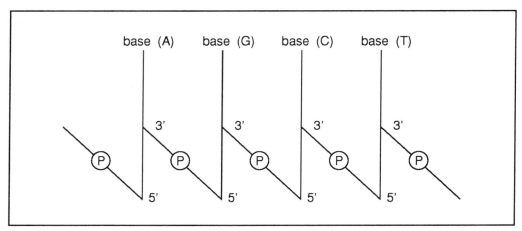

Figure 6.7 Shorthand notation to show the base sequence of a nucleic acid.

For most forms of RNA the simple relationship of A/U = G/C = 1 does not hold since the RNA molecule consists of a single strand of nucleic acid in the form of a random coil with only limited regions of base-pairing.

In contrast, DNA consists of two chains of polynucleotides interwoven in the form of a spiral structure which is stabilised by hydrogen bonding between particular base pairs.

6.3 The structure of DNA

SAQ 6.1

Although DNA and RNA share the same basic structure there are a few differences in terms of their 'building blocks'. Make a list of these differences.

complementary bases double helix

You should be aware that specific pairs of bases can form hydrogen bonds with each other. Figure 6.8 shows the pairings that are possible in DNA. Because of this hydrogen bonding between 'complementary bases', two strands of DNA can associate to form a 'double helix' structure (Figure 6.9).

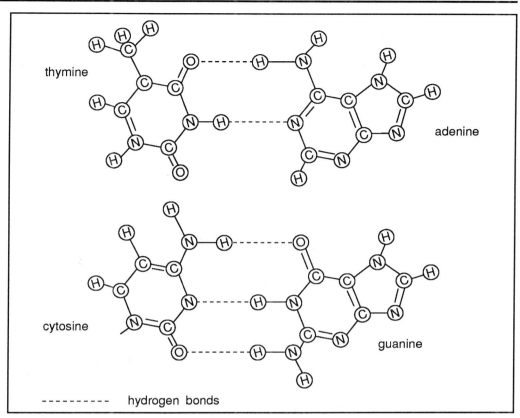

Figure 6.8 Pairing of complementary bases via hydrogen bonds.

Π Do you think that any two DNA strands could associated to form a double helix?

Because thymine will only pair with adenine and cytosine with guanine the entire sequence of one strand must be complementary to that of its partner if a perfect double helix is to result. The geometry of this structure is such that the two strands run in opposite directions (ie, one runs 5′ to 3′ and the other 3′ to 5′); the strands are said to be **anti-parallel** anti-parallel. The paired bases lie at the centre of the double helix and are stacked on top of each other rather like a pile of plates. Although individual hydrogen bonds are weak the cumulative effect of all the hydrogen bonds in a long, double-stranded DNA molecule is to give the molecule great stability at room temperature. If you find this hard to understand, think of a zip fastener; the individual links are pretty weak, yet we all happily rely on their cumulative strength in a complete fastener.

We remind you that specific base-pairing has important biological consequences; it ensures that the nucleotide sequences of DNA are conserved during replication and it also ensures that nucleotides are incorporated in the correct sequence during the synthesis of RNA. We will not however enlarge on these biological functions here.

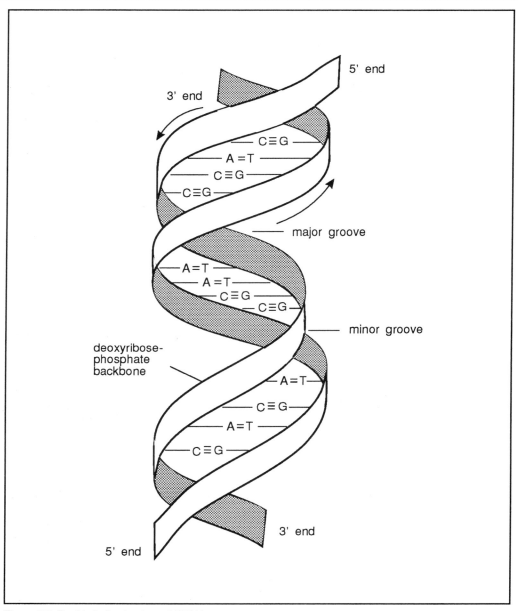

Figure 6.9 Double helix structure of (DNA)

<table>
<tr><td>SAQ 6.2</td><td></td></tr>
</table>

SAQ 6.2

1) List the bases found in RNA and DNA.

2) What other differences are there between RNA and DNA?

3) State the base pairing rules and explain why these are not obeyed in RNA.

6.4 The function of RNA

A thorough treatment of the function of RNA is beyond the scope of this chapter. Full details will be found in the BIOTOL text, 'The Infrastructure and Activities of Cells'. To enable you to understand some of the different types of RNA mentioned in this and later chapters, the following is a brief revision in the form of a list of the more important types of RNA together with their functions:

- messenger RNA (mRNA); information-carrying intermediates in protein synthesis;

- ribosomal RNA (rRNA); a component of ribosomes;

- transfer RNA (tRNA); amino acid delivery to the ribosome;

- small nuclear RNA (snRNA); components of the RNA splicing ribonucleoprotein particles;

- ribozymes; catalytic RNA molecules;

- enzyme component; eg, the 7 SLRNA of the signal recognition peptide;

- genomic RNA; double or single-stranded RNA genomes found in many viruses;

- primers for DNA synthesis.

The important thing to realise is that RNA molecules have varied functions, principally concerned with protein synthesis and that much has been learned about this biochemical process by isolating, purifying and studying RNA. Our understanding of the control of gene expression has also been assisted by studying RNA synthesis and many genes have been cloned by first isolating specific mRNA.

6.5 The principles and application of hybridisation

Molecular hybridisation is such an important technique used in the study of nucleic acids that it is essential that you understand the principles of this procedure.

We have already learnt that the bases in nucleic acids will tend to form hydrogen bonds with other bases. In principle, if we incubate single-stranded DNA (DNA in which the two strands of the double helix have been separated) in conditions in which hydrogen bonds can form, then the separate strands can combine through base-pairing providing that the two strands are complementary. This process can occur between two strands of DNA or between a strand of DNA and a complementary strand of RNA. Use can be made of this process to help us to identify nucleic acids with particular nucleotide sequences. For example let us assume that we have separated fragments of DNA by electrophoresis and 'blotted' them onto a membrane as described in Chapter 1. Suppose we now 'melt' (separate the strands) the DNA on the membrane and then challenge these resulting single-strands with another sample of single-stranded DNA. If this sample contains complementary nucleotide sequences to those of the DNA on the membrane, hydrogen bonds will be formed and a hybrid double-stranded DNA molecule will be formed. We are effectively using base-pairing as a method of identifying nucleic acids containing particular nucleotide sequence. Usually the challenging nucleic acid is labelled (for example, made radioactive) so that it can be detected and measured by autoradiography or by some other technique. More recently non-radioactive labelling has become more important. A common label used is biotin chemically attached to the challenge nucleic acid (often called a probe). Biotin is very tightly bound by a protein called avidin. If we attach an enzyme to the avidin then we can detect where avidin binds by measuring the enzyme. We can represent the sequence of events as follows:

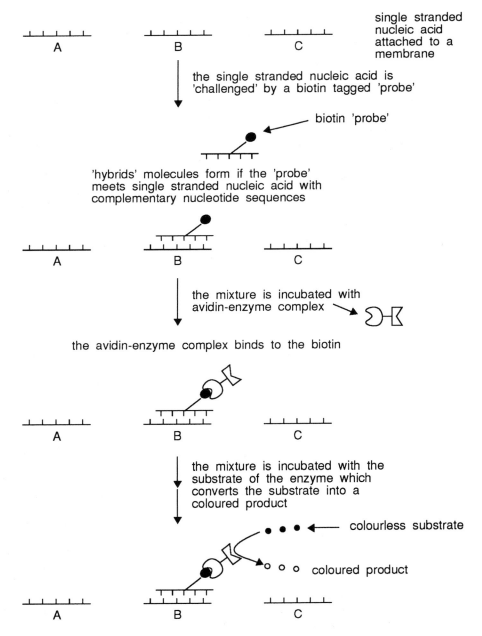

You should note that there are many variations of this general theme.

Π Is the 'probe' used in the flow diagram above specific for a particular nucleotide sequence?

The answer is yes, it only binds with the sequence in molecule B. We can use the amount of enzyme reaction in the final stage to measure the amount of hybridisation that has taken place, which, in turn, may tell us how much of sequence B was present. Hybridisation can therefore be used as both a qualitative (to identify if a particular sequence is present) and as a quantitative (how much of a particular sequence is present) technique. The technique of blotting DNA from electrophoretograms and using DNA hybridisation to detect the DNA on the membranes is called Southern blotting. A similar technique, rather coyly called Northern blotting, can be used to detect RNA separated by electrophoresis.

From the description of the technique given above it may appear simple. However, there are many pitfalls. One of these is that if we choose conditions in which hydrogen bonds are very stable (for example at a low temperature), then hybrids may form in which there are many mismatched base pairs. This of course will lower the specificity of the technique. We will deal with mismatching and the 'stringency' of the hybridisation conditions when we discuss the analysis of DNA and RNA in later chapters. For now, the important message is that we can use single-stranded nucleic acids to detect and identify nucleic acids containing particular nucleotide sequences.

SAQ 6.3

1) A bacterial genome is cut into fragments and the individual fragments separated by gel electrophoresis. After blotting onto a cellulose nitrate membrane, the DNA was denatured to form single stranded DNA. This DNA was challenged with ^{32}P-labelled DNA prepared from a bacteriophage which contained single stranded DNA. After incubation, the membrane was washed and placed against an X-ray film. After further incubation, the X-ray film was developed and revealed 2 blackened areas. What can be concluded from this experiment?

2) A similar electrophoretogram was incubated with ^{32}P-mRNA which codes for the enzyme β-galactosidase. After washing, an autoradiogram was produced which had one blackened area. What can be concluded from this experiment?

6.6 Gene engineering and the analysis of nucleic acids

It is perhaps a paradox that the study of nucleic acids is essential to genetic engineering and that the genetic engineering has facilitated the study of nucleic acids. A common problem encountered in nucleic acid research is that often the researcher has to carry out analysis on very small quantities of the nucleic acids. Genetic engineering enables researchers to make much larger quantities of the nucleic acid of interest. Here we will outline the steps involved in genetic engineering so that you will understand its importance to nucleic acid analysis. It is not our intention here to go into a lot of technical details, these are more appropriate to discussion of genetic engineering *per se*. The techniques of genetic engineering and the outcomes of the application of these techniques are developed in the BIOTOL texts, 'Techniques for Engineering Genes', and 'Strategies for Engineering Organisms'.

6.6.1 The major steps in genetic engineering

genetic
engineering

The process of genetic engineering is conveniently described in four stages. These are:

- the preparation of the gene which is to be transferred;

- the incorporation of the gene into a suitable vector;

- the introduction of the vector into the new host cell (a process called transformation);

- demonstration that successful incorporation of the gene has taken place and that the product coded by the gene is being made (gene expression).

Before considering each of these stages, the properties of restriction enzymes will be considered.

Restriction enzymes are enzymes which hydrolyse double-stranded DNA at particular points in the DNA. Each restriction enzyme (several hundred have been described) recognises a particular sequence of bases in DNA and only cuts the DNA at this site. For example, the restriction enzyme Eco R1 (obtained from *E. coli*) recognises the sequence GAATTC and only cleaves at this sequence. Use of restriction enzymes enables DNA to be reproducibly cut at a relatively few sites (since each recognition sequence only occurs infrequently). Restriction enzymes are essential for the production of particular genes, and are also indispensable in other areas of applied molecular biology (DNA sequence analysis, DNA fingerprinting, DNA probes).

6.6.2 Preparation of the gene

sticky ends

Bacterial genes can usually be obtained by digesting chromosomal DNA preparations with restriction enzymes. In most cases the restriction enzyme used generates fragments with overlapping (or sticky) complementary single-strand ends (Figure 6.10).

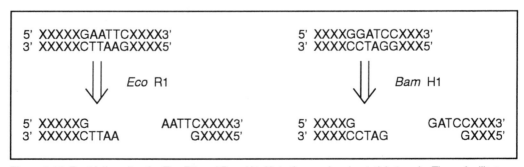

Figure 6.10 Restriction sites for Eco R1 and Bam H1. Note the overlapping (sticky) ends. These facilitate joining of pieces of DNA, providing their ends are complementary.

Genes coding for small gene products (eg peptide hormones may be chemically synthesised, providing the amino acid or DNA sequence is known. Eukaryotic genes (eg those from humans) cannot be produced in a form which is suitable for genetic engineering by restriction enzyme digestion of chromosomal, DNA, because of the non-coding regions (introns) which they contain. These have to be removed before protein synthesis takes place. As bacterial genes do not contain introns, bacteria have no system for recognising and removing introns after transcription. Unless the introns present in eukaryotic DNA are removed prior to introduction into a bacterial host, a protein with incorrect amino acid sequence will be produced. A gene which will be correctly expressed in bacteria is produced by making a DNA copy (cDNA) of the mRNA present in eukaryote cells (Figure 6.11); this cDNA is used for genetic engineering. If you examine this figure carefully, you will see that by making a DNA copy of the mRNA isolated from eukaryotes, we can produce a new version of the gene without intron sequences. The key enzyme in this process is reverse transcriptase. This uses RNA as a template and produces a DNA copy of this. The enzyme can be more properly called RNA-dependent DNA polymerase.

cDNA

You should release that we do not have to use an intact gene if we are merely interested in making copies of a particular nucleotide sequence.

We also talk about 'libraries' of genetic information. If for example we use a whole genome as our source of genetic information and fragment this using a restriction enzyme; when we have cloned this into a vector (see Section 6.6.3) we should end up with a family of vectors each carrying different fragments of the genome. In other words we would have a 'library' of genomic fragments. The collection of various 'cloned' fragments of the genome is called a genomic library.

genomic library

Similarly, if we begin with a mixture of all of the different mRNAs produced by a population of cells and convert these to DNA sequences using reverse transcriptase, we can produce what is called a cDNA library.

cDNA library

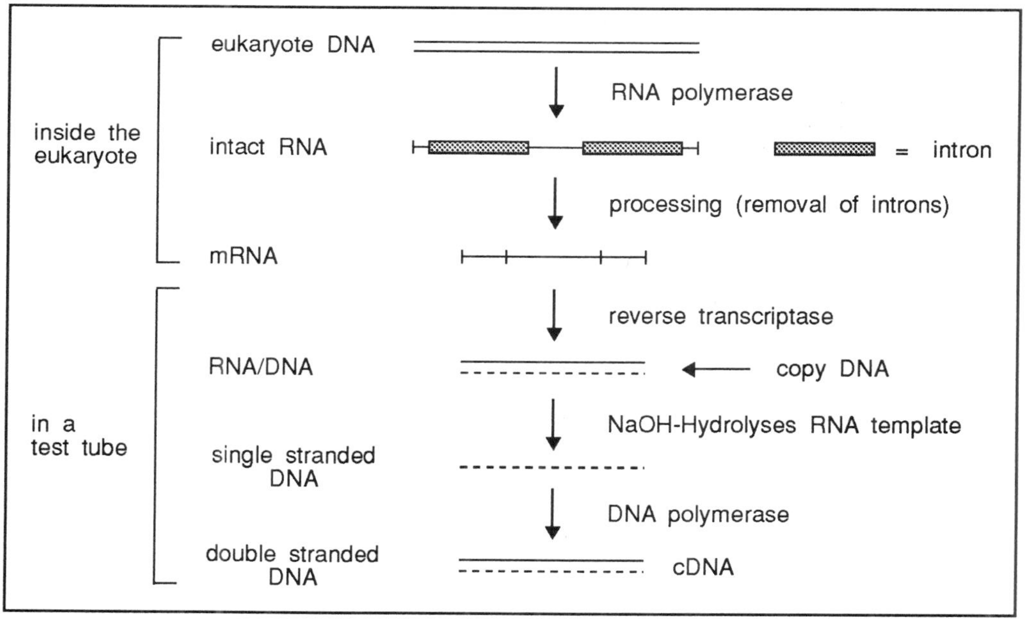

Figure 6.11 Preparation of cDNA (in outline). The processed mRNA (which now codes for the desired protein) is used as a template by reverse transcriptase to produce a single stranded copy DNA. This can be used to generate double stranded cDNA.

6.6.3 Incorporation of the gene into a vector

vector

plasmids

In order to ensure that the foreign gene is taken up and expressed by the host cell, a vector is required. Foreign DNA is not normally incorporated into the bacterial chromosome but must none-the-less divide (replicate) when the host cell divides. The vector has to act as a replicon in order that the novel gene is maintained in the host cell. Very often so called plasmids are used as vectors. Plasmids are circular pieces of double-stranded DNA which are replicated within the host cell. Useful features of plasmids used for genetic engineering include the following:

- single sites for several restriction enzymes;

- presence in multiple copies within the host cell (thus one obtains many copies of the novel gene);

- plasmid codes for selectable markers, such as antibiotic resistance: this permits selection of transformants.

Widely used plasmids, such as pBR322 (Figure 6.12), contain these features.

For use, the vector is cut with the same restriction enzyme as that used to produce the gene for insertion: thus gene and vector will have complementary sticky ends (Figure 6.13).

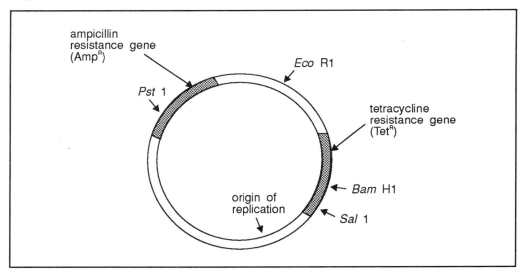

Figure 6.12 pBR322. The positions of certain restriction enzyme sites, which only occur once in the plasmid, are indicated. Also shown are the two genes whose products enable selection of transformed *E. coli* cells to be made. This type of diagram in which the position of restriction sites is marked, is known as a restriction map.

The prepared gene and vector are incubated with DNA ligase, which forms covalent bonds where complementary ends overlap.

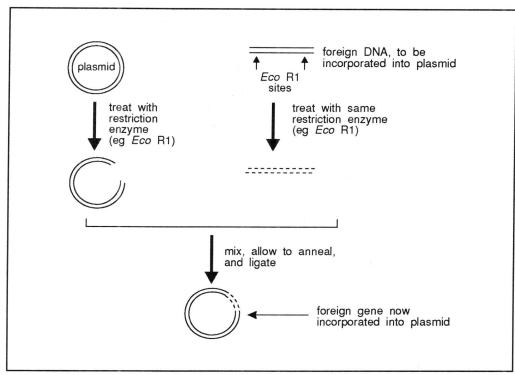

Figure 6.13 Introduction of foreign DNA into a vector. The plasmid and foreign DNA are cleaved with the same restriction enzyme: this ensures that they both have complementary sticky ends (note that there are other ways of achieving complementary ends). Upon mixing, some of the fragments will re-anneal to give a plasmid containing an insert; DNA ligase forms the necessary covalent bonds.

∏   In Figure 6.13 we use the word anneal. What do you think we mean by this term?

The term anneal used in this context means that hydrogen bonds form between complementary sequences of nucleotides to give a double-stranded structure. If you look back to Figure 6.10, you will see that the overlapping ends of restriction enzyme fragments are complementary. Thus if we incubate two fragments together under conditions in which hydrogen bonds can form, the two fragments will join together. This process of hydrogen bond formation between the two strands is often referred to as annealing.

6.6.4 Transformation of host cells

A suitable vector containing the inserted gene is then introduced into a bacterial host cell. When *E. coli* is the host, treatment with calcium chloride results in easily transformed cells. Transformed cells may be selected by their ability to grow in the presence of an antibiotic, resistance against which is coded by the vector (eg, for *E. coli* transformed with pBR322, the plasmid confers resistance to the antibiotics ampicillin and tetracycline). Selection by antibiotic resistance is highly effective (Figure 6.14).

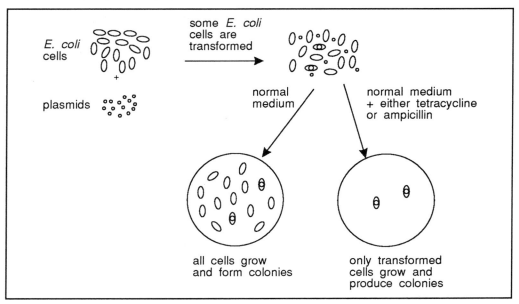

Figure 6.14 Selection of transformed cells. *E.coli* cells lacking pBR322 fail to grow on medium containing either ampicillin or tetracyclin. The presence of the plasmid confers resistance on the *E. coli* cells.

This means that only a small proportion of bacteria present need to be transformed since they are easily selected.

⫪ Assume that some foreign DNA is incorporated into the single *Bam* H1 restriction site of pBR322, which is then used to transform cells of *E. coli*. Predict the phenotype of the transformed cells with regard to the antibiotics ampicillin and tetrancycline. (By phenotype we mean the characteristics displayed by the organism; in this case, are they sensitive to or resistant to these antibiotics?).

The phenotype will be Amp^R, Tet^S, meaning that the *E. coli* cells are resistant to ampicillin but sensitive to tetracycline. Look at the restriction map of pBR322 (Figure 6.12). Successful incorporation of extra DNA into the *Bam* H1 site of pBR322 means that, although the tetracycline gene will still be transcribed, it will produce a protein with an extra section in the middle of it. It is most unlikely to be functional; this is known as insertional inactivation. Thus the gene product conferring resistance to tetracycline is not produced in a functional form. Hence Tet^S. The penicillinase which confers resistance against ampicillin is not affected, hence the transformed cells would be Amp^R.

insertional
inactivation

Note that if foreign DNA was not incorporated into pBR322 and it simply re-annealed without an insert (eg in Figure 6.13, intact pBR322 is reformed), or if the pBR322 was not cut by the *Bam* H1 at all, then the phenotype would be Amp^R, Tet^R. *E. coli* cells which are not transformed will be Amp^S, Tet^S. These phenotypes can all be distinguished by plating onto appropriate media. Incorporation of foreign DNA into the Pst 1 site of pBR322 would give an Amp^S Tet^R phenotype. These insertional inactivations are very useful as a means of monitoring the success of incorporation of foreign DNA and transformation.

6.6.5 Detection of the cloned gene and gene product

in situ
hybridisation

The presence of the gene within transformed cells may be demonstrated by the use of a radioactively labelled DNA probe which, by complementary base-pairing, only binds to the complementary sequence (Figure 6.15).

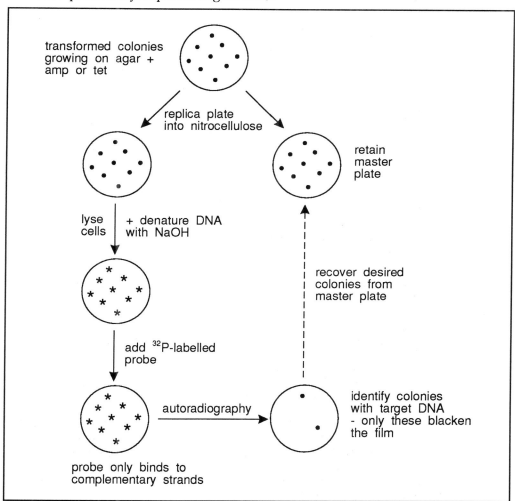

Figure 6.15 Colony hybridisation (also called *in situ* hybridisation). DNA from transformed cells is denatured and bound to a nitrocellulose support. 32-labelled DNA (the probe) only binds to complementary sequences. Use of an appropriate probe reveals which colonies contain the desired DNA sequence. Adapted from Old, R.W. and Primrose, S.B. (1985) 'Principles of Gene Manipulation,' 3rd Edition, Blackwell.

This enables colonies of cells which contain the gene or desired nucleotide sequence to be identified. We can of course then grow large quantities of the cells carrying the nucleotide sequences of interest. We can extract the DNA (plasmids) from these cells, fragment it using restriction enzymes and then purify and identify the fragments(s) of interest by Southern blotting. In this way we can produce a large quantity of the nucleotide sequence of interest. We will deal with the isolation of plasmids in Chapter 7. There is however another way of producing greater quantities of desired nucleotide sequences. This uses an *in vitro* system called the polymerase chain reaction (PCR) technique.

SAQ 6.4

1) Which of the following nucleotide fragments are most likely to anneal with each other?

   a)   AATTGATTAA   b)   AATTGCTAG
             TTAACT   TTAAC

   c)   AATTGTTAA   d)   AATGCATG
             TTAAC   TTAC

2) Is a genomic library more or less likely to contain a particular gene than a cDNA library?

3) List at least two circumstances in which using a cDNA library is unlikely to lead to the successful cloning of a desired nucleotide sequence.

6.7 The polymerase chain reaction (PCR) technique

The polymerase chain reaction technique allows the specific amplification of DNA. The method is based on the repetition of a set of three steps all conducted in succession under different and controlled temperature conditions. Use Figure 6.16 to help you follow the description given. Each cycle consists of the following steps:

- Denaturation: the double-stranded DNA sample is denatured by incubation at high temperature ($90°C$).

- Annealing of extension primers: the extension primers constitute a pair of synthetic oligonucleotides, not complementary to each other, but to both ends of the DNA sequence that must be amplified. After cooling to $55°C$ the primers will anneal to the DNA in the sample at the boundaries of the region to be amplified. The primers anneal to opposite strands with their 3' ends facing each other. The primers are added in a large excess over the DNA template and thus primer template complex formation is favoured over reassociation of the two DNA strands when the temperature is lowered.

- Primer extension: the DNA polymerase extension ($5' \rightarrow 3'$) of the primer template complex is carried out at $70°C$ with a thermostable DNA polymerase purified from *Thermus aquaticus* (Taq DNA polymerase). During the succession of cycles the amount of product derived directly from the template molecules in each cycle (the 'long product') will increase arithmetically because the quantity of original template remains constant. The product of interest in the amplification product is the region comprised between the 5' ends of the extension primers (the 'short product'). Since the primers have well defined sequences, the short product will have well defined ends, corresponding to the primer sequences. As the number of cycles increases, the short product will rapidly become the predominant template to which the extension primers will anneal, which leads (theoretically) to an exponential accumulation. The efficiency found in practise is usually somewhat lower. The PCR method has been fully automated.

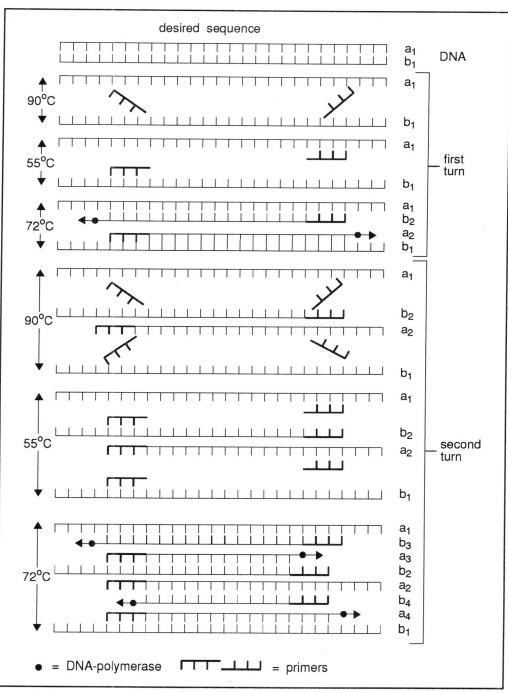

Figure 6.16 The polymerase chain reaction (see text for a description).

Summary and objectives

In this chapter we briefly reviewed some of the chemical and physical properties that you will need to know in order to understand the discussion of nucleic and extraction and purification in subsequent chapters. We have also explained that single stranded nucleic acids can form hydrogen bonds with other strands providing they carry complementary nucleotide sequences. This process, referred to as hybridisation is of great practical importance for the detection and measurement of particular nucleotide sequences. We have also learnt that the process of hydrogen bond formation between complementary sequences is important in genetic engineering and the polymerase chain reaction technique. These latter two topics were described as procedures which allow us to make larger quantities of desired nucleotide sequences and, therefore, of great value to the researcher.

Now that you have completed this chapter you should be able to:

- list the main differences between RNA and DNA;

- describe what is meant by hybridisation as applied to nucleic acids and explain how hybridisation may be used to detect and measure specific nucleotide sequences;

- draw an outline scheme for producing a large quantity of a particular nucleotide sequence using a genetic engineering approach;

- draw the sequence of events for producing desired nucleotide sequences using the polymerase chain reaction technique.

Extraction and estimation of deoxyribonucleic acid (DNA)

Extraction and estimation of deoxyribonucleic acid (DNA)

7.1 Introduction

This chapter begins with a brief survey of the potential reasons for extracting DNA. Extraction of DNA is discussed in terms of starting tissue, homogenisation and removal of major contaminants. Strategies for DNA isolation from whole cells, nuclei, cytoplasmic organelles, bacteria and viruses are outlined.

Before considering methods for the extraction of DNA we should have in mind the possible reasons for doing such extractions. Since DNA carries the genetic information of prokaryotic and eukaryotic cells (including their organelles) and of many viruses, its structure, particularly its sequence of nucleotides, is of fundamental importance to living organisms. Its importance to biological scientists is well illustrated by the 'Human Genome Project', which has as its aim the discovery of the entire nucleotide sequence of the DNA in the human nucleus. This massive project is expected to take at least ten years to complete and involves the concentrated efforts of many large research teams. It has only become feasible with the development of new methods for the analysis of DNA. The computing problems posed by the generation of so much data are another story!

We need to extract DNA not only to sequence it but also if we want to modify it using the methods of genetic engineering. It is no exaggeration to say that genetic engineering is transforming biological and medical sciences and you are referred to the BIOTOL texts, 'Techniques for Engineering Genes,' and 'Strategies for Engineering Organisms' for details of this exciting technology.

For most purposes the DNA we isolate must be relatively pure and should not be damaged physically. It is also important to be able to determine the purity and quantity of our samples. All these aspects of DNA manipulations are dealt with in this chapter.

7.2 Extraction of DNA

7.2.1 Choice of tissue

One of the most important decisions to be made when designing a DNA extraction procedure concerns the choice of starting material.

∏   What are the main properties you would look for in your starting material?

You probably realised that it is important to choose a tissue that has a high content of DNA. In general this implies the presence of a large number of small cells per unit volume and, consequently, a high concentration of nuclei. This rule might not apply if you are aiming at the extraction of organelle DNA, where the concentration of organelles rather than nuclei will be of more importance. If you consider that the tissue must be broken open in order to release its DNA you will also realise that a soft, fragile tissue is very desirable.

| SAQ 7.1 |

Which of the following would be suitable for the isolation of nuclear DNA?

a) liver; b) bone; c) red blood cells; d) leaf buds; e) potato tubers.

shearing forces

The moment we start breaking open cells and nuclei the DNA is removed from its protective environment. What are the dangers facing it? Although DNA is often thought of as a long, flexible molecule which can readily fold and bend, it behaves in solution as if fairly brittle and is, therefore, easily snapped by mechanical shearing forces. Such forces occur when adjacent layers in a solution are moving at different speeds. For example, when liquid flows through a tube the liquid in contact with the tube wall is slowed down whilst that at the centre of the tube is relatively free to flow. If the tube is narrow and the liquid is flowing rapidly, shearing forces can be considerable - certainly enough to break long DNA molecules into shorter fragments.

∏ What extraction processes are likely to generate shearing forces?

We do not usually force DNA solutions through narrow tubes during extractions, but rapid flow through narrow holes in pipette tips can damage high molecular mass DNA. If high molecular mass DNA is removed from a CsCl gradient (Section 1.2.2) using a syringe the needle should be wide bore and flow rate should be kept as low as possible. Too much vigour during homogenisation or when mixing DNA solutions with alcohol or phenol can damage DNA, so such processes should be thorough yet gentle.

deoxy-
ribonucleases

chelating agent
EDTA

Another major hazard to DNA is the release of deoxyribonucleases (DNases) from lysosomes, vacuoles, etc into the extraction medium. Fortunately most DNases require divalent cations, most usually Mg^{2+}, for activity and so can be inactivated by the presence of a chelating agent, such as EDTA, which binds divalent cations strongly. Since DNases could, potentially, be introduced in the medium or be present on glassware, all extraction media and glassware are autoclaved before use to inactivate DNases.

7.2.2 Homogenisation (see also Section 1.1)

To extract total DNA from a tissue we must disrupt the tissue and break open the cells. Soft animal tissues, such as liver, can be readily disrupted mechanically by very brief homogenisation using a blender (Section 1.1.1) or Polytron. Such treatments will break many cells and will, inevitably, cause some physical damage to organelles and some shearing of DNA. Every homogenisation is, therefore, a compromise; vigorous homogenisation will give high yields of damaged, contaminated DNA, gentle homogenisation will give low yields of high quality DNA.

The main steps in DNA extraction from eukaryotes are:

• preparation of tissue;

• disruption of tissue;

• lysis of cells.

Subsequent steps depend on the type of DNA required, as shown in Figure 7.1.

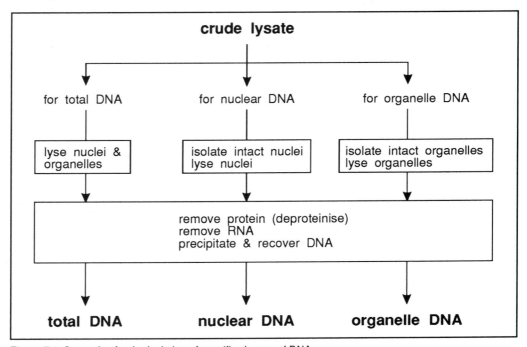

Figure 7.1 Strategies for the isolation of specific classes of DNA.

mitochondrial
DNA
chloroplast
DNA
nuclear DNA

The total DNA of eukaryotes will include some mitochondrial DNA (mt DNA) and, in the case of plants, chloroplast DNA (cp DNA). Since the latter can form up to 20% of the total DNA it is not wise to regard total DNA as exclusively nuclear DNA (nDNA). Thus many DNA extractions begin with the isolation of nuclei or organelles.

We shall not deal in detail with nuclei and organelle isolation, which is discussed in the BIOTOL text, 'Techniques used in Bioproduct Analysis'. However, most procedures rely on differential centrifugation (Section 1.2.1) to produce samples enriched in the desired organelle, followed by density gradient centrifugation to give a purer preparation.

Π All membranes are based on a lipid bilayer. What sort of reagent would you expect to solubilise membranes and so lyse nuclei and organelles?

You only have to think of washing dishes to realise that detergents solubilise lipids.

lysis with
detergents

Commonly used detergents include sodium dodecyl sulphate (SDS) and Sarkosyl. The non-ionic detergent Triton X-100 does not dissolve nuclear membranes but will lyse organelles.

Π How could this property be exploited?

A preparation of nuclei contaminated with mitochondria and chloroplasts can be treated with Triton X-100 to lyse the organelles. Intact, pure nuclei can then be collected by gentle centrifugation for subsequent lysis using SDS, thus releasing pure nDNA.

Since prokaryotes have no nucleus or organelles their DNA is obtained by lysis of the cell membrane using detergent.

∏ Do you think this treatment will be sufficient to lyse bacteria?

lysozyme

If you remember that bacteria have tough cell walls you will realise that we also need to break these walls. With Gram-negative bacteria such as *Escherichia coli* this can be done using the enzyme lysozyme, which cleaves the peptidoglycan backbone of the cell wall.

plasmids

In addition to their main, chromosomal DNA, bacteria may contain small circles of DNA known as plasmids. These carry a limited amount of genetic information, such as genes for resistance to specific antibiotics, and have been engineered by molecular biologists to make them very useful in the cloning of fragments of 'foreign' DNA. Since plasmids are used in genetic engineering it should be apparent that we frequently need to recover pure plasmid from a bacterial culture.

cleared lysate method

There are several methods for plasmid isolation but they tend to have some common features. The cells are usually grown in liquid culture, harvested by centrifugation and re-suspended in a small volume of buffer solution containing EDTA and an osmotically active compound such as sucrose. The cell wall is then weakened by digestion with lysozyme and the cell membranes are lysed by detergent. The small plasmid molecules escape into the buffer solution but chromosomal DNA is so long that it becomes tangled and forms a gel-like mass which can be sedimented by high speed centrifugation (hence this is often termed the cleared lysate method of plasmid isolation). The supernatant is, therefore, enriched in plasmid. Standard RNase, deproteinisation and ethanol precipitation are then used to purify the plasmid (we will discuss these in more detail a little later). Variations on this theme include the use of heat or alkali to denature DNA; on cooling or neutralisation only the plasmid DNA renatures, making it easy to sediment and remove the denatured chromosomal DNA.

∏ Why is sucrose included in the isolation medium?

What do you think would happen to a bacterium if it was treated with lysozyme in the absence of an osmotically active compound? As its cell wall lost its rigidity the cell would take up water and swell rapidly, eventually bursting violently and so damaging the chromosomal DNA by shearing. Since the separation of plasmid from chromosomal DNA relies on size differences such damage would cause contamination of the plasmid.

∏ Can you remember what most viruses consist of? Does this suggest how you could isolate viral DNA?

viral DNA

Most viruses consist of a core of nucleic acid (often DNA) surrounded by a coat of protein. Once virus particles have been separated from other material (usually by differential centrifugation) a simple deproteinisation is all that is required to liberate pure viral DNA.

SAQ 7.2	List the main steps you would use to prepare an extract of chloroplast DNA from peas.

7.2.3 Further purification of DNA

The next steps in the purification of DNA are the removal of protein and RNA. The deproteinisation of extracts can be undertaken in a number of ways but the most widely used method is extraction with phenol or a phenol/chloroform mixture. In this method, the proteins are removed by mixing the lysate with an organic solvent, usually phenol or phenol/chloroform mixture. What do you think will happen when proteins in the homogenate come into contact with phenol?

The second and tertiary structure of proteins and polypeptides is highly dependent on the chemical environment. This is usually aqueous in nature with polar amino acids on the outside of the protein. The presence of an organic solvent will denature the proteins by disrupting the hydrogen bonding.

∏   What happens to a solution of protein when the molecules are denatured?

Think of cooking egg whites (a solution of ovalbumin and other proteins) at home. The polypeptide chains unfold and the protein precipitates; in the case of egg white into a solid mass. Phenol will also bring about this. Centrifugation of the homogenate-phenol mixture will, therefore, give rise to three phases (Figure 7.2).

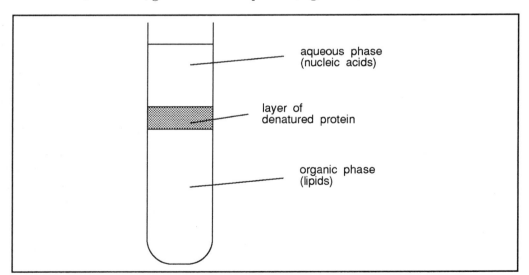

Figure 7.2 Separation of soluble nucleic acids and denatured protein by phenol/detergent extraction.

phenol
extracton

The nucleic acids will be dissolved in the upper layer and thus will be separated from proteins that form a solid mass in the middle, and lipids, dissolved in the lower (phenol) organic phase. The aqueous layer is simply removed with a pipette; it is common practice to re-extract it with a further equal volume of phenol mixture to ensure that all proteins have been removed. The aqueous phase at this point is virtually a pure nucleic acid solution; the principal contaminants are low molecular mass compounds and possibly carbohydrate, depending on the starting material, and of course, detergent if this was used in the extraction.

Before carrying out the centrifugation remember that you must shake the sample sufficiently to generate an emulsion but not so violently that the DNA is sheared. Depending on the level of protein in the extract you may have to do several cycles of deproteinisation to remove all the protein. Although the aqueous solution containing DNA and the organic phenolic phase separate from each other during centrifugation some phenol partitions into the aqueous phase. Even traces of phenol can inhibit certain enzymes that might subsequently be used with your DNA sample and so it is worth removing these traces by a final extraction with chloroform or ether.

proteinase K

Sometimes certain proteins remain stubbornly bound to the DNA (this is frequently seen with the histone proteins of nuclear DNA) and are not removed by phenol. In such cases treatment with a proteolytic enzyme, usually Proteinase K, is required to digest the proteins.

Often EDTA (ethylene diamine tetra-acetic acid) is included in the homogenisation buffer. EDTA chelates divalent metal ions that are essential for DNase activity. Inclusion of EDTA in the extraction medium generally completely inhibits DNase and therefore protects the DNA from enzymatic degradation.

∏ After centrifugation, what will be the major components of your (aqueous) extract?

There should, of course, be DNA in the extract. There will also be RNA and low molecular mass molecules such as amino acids, nucleotides and sugars. Depending on the source of your extract there could also be significant levels of carbohydrate.

RNA can easily be digested by the addition of RNase. Since commercial preparations of RNase are often contaminated with DNases it is usual to make up a stock solution of RNase and then incubate it at 100°C for 5 minutes to denature the enzymes. On cooling, DNases remain denatured but RNase renatures readily because of the -S-S- bonds holding it together. Such heat-treated stocks are safe to use for DNA preparation.

So how is the DNA to be recovered from the extract? The simplest thing to do is to exploit the high molecular mass and polar nature of DNA. DNA is polar and will precipitate in appropriate organic compounds. Ethanol is added (approximately 2.5 times the volume of the aqueous phase). For good yields of DNA the solution is made approximately 0.3 mol l^{-1} with respect to sodium acetate to form a DNA salt and then two volumes of chilled ethanol are added to precipitate the DNA (remember to mix well!). After a few minutes on ice or in a -20°C freezer the preparation should be centrifuged for at least ten minutes to sediment DNA.

∏ What molecules would you expect to find in the supernatant?

Molecules of low molecular mass, including the ribonucleotides produced by RNase treatment, should remain in the supernatant. The only contaminant of the DNA pellet should be carbohydrate. High levels of carbohydrate could inhibit some enzymes used in molecular biology. If this is a problem you can remove carbohydrate by extracting the aqueous solution with 2-methoxyethanol.

Finally, after a quick wash with 70% ethanol, the DNA pellet can be re-dissolved, usually in 'TE' (10 mmol l^{-1} Tris-HC1, pH 8.0, 1 mmol l^{-1} EDTA).

∏ What is the function of EDTA in the TE?

By chelating Mg^{2+} the EDTA ensures that any DNase which might contaminate the DNA solution will be inactive. Pure DNA can be stored in TE at 4°C for several weeks.

7.3 Measurement and estimation of DNA

Estimation using UV spectroscopy

Nucleic acids have a marked peak absorbance in the ultra-violet region at 260 nm (Figure 7.3).

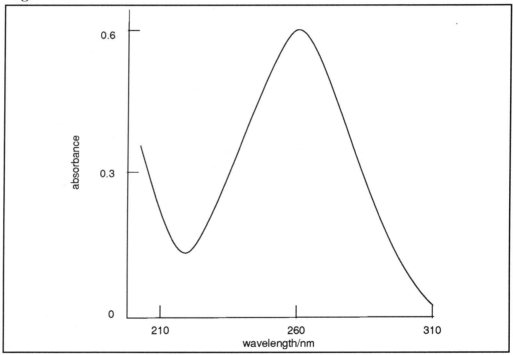

Figure 7.3 Ultra-violet absorption spectrum of a nucleic acid solution.

Proteins which also absorb in the UV region have a peak at 280 nm. Thus pure nucleic acid solutions show an absorbance ratio A_{260}/A_{280} of 1.8 to 2.0 (pure DNA should give the lower figure, and RNA the higher, although this slight difference in ratio is inadequate as a means of distinguishing the two forms of nucleic acid). The value of peak absorbance at 260 nm can be used to calculate the nucleic acid concentration; for RNA an absorbance of 1.0 is approximately that of a solution of concentration 45 µg cm^{-3}. The figure is approximate because the degree of secondary structure of RNA affects the absorbance. The main point of difference when comparing DNA with RNA is that an absorbance at 260 nm of 1.0 indicates a concentration of approximately 50 µg cm^{-3} DNA assuming the solution contains only DNA. As little as 0.5 µg cm^{-3} of

DNA can be detected by UV absorbance, but because RNA and other contaminants will contribute to the absorbance at 260 nm this method cannot be used as anything more than a guide to the maximum DNA concentration.

If protein contamination is present the ration A_{260}/A_{280} will fall below 1.8 and the peak will shift from 260 nm. Other contaminants may also interfere, eg phenol and carbohydrates absorb in the same range of wavelengths. If is often necessary to remove contaminating carbohydrates (from RNA preparations by centrifugation of RNA in sodium acetate, or from DNA solutions by CsCl density gradient centrifugation) (Section 1.2.3).

Estimation of DNA by the Burton assay

Burton assay

For more accurate measurements of DNA concentration in impure samples it is possible to use the Burton assay. This assay is based on the hydrolysis of DNA with perchloric acid (usually about 0.5 mol I^{-1} perchloric acid at 70°C for 15 minutes) and subsequent reaction of diphenylamine with the hydrolysis products to produce a coloured product. This assay is specific for DNA rather than RNA but is not very sensitive, requiring at least 5 μg cm^{-3} DNA. Note that the perchloric acid hydrolysis only removes the purine residues of DNA.

Estimation of DNA using fluorescence

fluorescence

An alternative colorimetric assay, based on fluorescence, involves the removal of purine bases from DNA using acid. The resulting, depurinated DNA is then allowed to react with diaminobenzoic acid (DABA) to give a fluorescent compound. As little as 30 ng of DNA can be detected in the fluorometer cuvette.

ethidium
bromide

Ethidium bromide has the structure shown in Figure 9.4. You will see that it has a three-membered ring structure that, because of its conjugated double bonds, is flat. The molecule also absorbs ultraviolet radiation and re-emits some of the absorbed energy by fluorescence, to give an orange light.

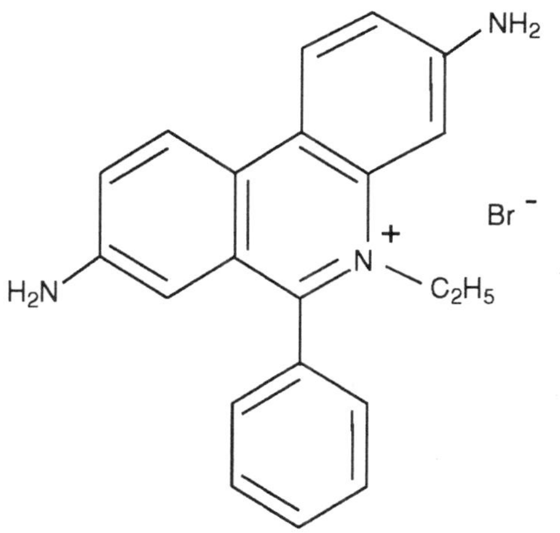

Figure 7.4 Structure of ethidium bromide.

Π This molecule binds to DNA. How do you think it might do this and how could we make use of this binding?

intercalation

A flat molecule like ethidium bromide would not bind tightly to the curving outer surface of the DNA double helix. However, it can slot in between adjacent stacked base pairs at the centre of the molecule by a process known as intercalation.

When this happens the base pairs on either side of the ethidium bromide are forced to move apart from each other to make way for the inserted molecule. Consequently the DNA lengthens and unwinds very slightly for each molecule of ethidium bromide that intercalates. Because the ethidium fluoresces in the ultraviolet region it acts as a dye and so is widely used to reveal the location of DNA in electrophoresis gels and CsCl gradients.

Ethidium bromide can be used to quantify DNA in solution. Using a fluorometer a standard curve of fluorescence against DNA concentration can be constructed and used to determine the concentration of an unknown DNA solution. This is a very sensitive method but it is not completely specific for DNA since RNA will also bind the dye to some extent. Another problem arises from the discovery that circular molecules of DNA, such as plasmids, will not bind as much dye as linear molecules and so will fluoresce less than would be expected for a given concentration of DNA. It should be pointed out that ethidium bromide is a potent carcinogen (it interacts with DNA). Therefore great care must be taken in using this chemical and it is usually chemically destroyed before disposal. When you come to carry out these procedures in the laboratory you must make yourself familiar with the rules governing the use of this hazardous chemical.

For molecular biology it is usually sufficient to have an approximate value for the concentration of DNA in a sample. This can be obtained by running the sample on an electrophoresis gel (see Chapter 1) alongside a range of standard amounts of DNA. The gel is then stained with ethidium bromide to reveal bands of DNA and the intensity of fluorescence of the unknown is compared, by eye, with the standards. This procedure requires very little solution and is very sensitive (as little as 1 ng DNA will give a visible band). An added advantage is that the electrophoresis separates high molecular mass

---

SAQ 7.3

Which method would you use to estimate DNA in each of the following samples?

1) 100 cm^3 of crude extract from 20 g pea leaves.

2) 3 cm^3 of pure, fairly concentrated DNA.

3) $50 \, \mu l$ of total nucleic acids extracted from a blood stain.

DNA from RNA and so gives an indication of the purity of the sample.

Summary and objectives

Where possible, DNA is extracted from tissue that is soft and rich in DNA. The tissue is homogenised gently with EDTA in the extraction buffer solution. Contaminants are removed by RNase, deproteinisation and ethanol precipitation. Special procedures may be required for extraction of DNA from organelles, bacteria or viruses. DNA can be assayed by UV absorbance, Burton assay, DABA assay or ethidium bromide fluorescence; only the Burton and DABA assays are specific for DNA.

On completion of this chapter you should now be able to:

- describe the structure of DNA;

- choose an appropriate tissue for DNA extraction;

- choose an appropriate strategy for isolating DNA from a range of sources;

- describe some methods commonly used to purify DNA;

- explain and compare four methods for estimating DNA and select appropriate methods for estimating DNA in particular samples.

Purification of molecular species of DNA

Purification of molecular species of DNA

8.1 Introduction

This chapter begins with a consideration of the factors causing contamination of DNA preparations. The main methods for obtaining a pure species of DNA are then discussed, including preparative agarose gel electrophoresis and isopycnic ultracentrifugation.

In Chapter 7 we discussed the extraction of DNA from cells and the removal of contaminants such as RNA, proteins, lipids and carbohydrates. We also saw how to enrich the preparation for nuclear or organellar DNA.

∏   Can you remember how to prepare a sample enriched in mitochondrial DNA?

The essential step in organella DNA preparations is to purify the organelles as far as possible before lysing them to release their DNA.

It is important to realise that organelle preparations involving differential centrifugation are merely enriched for the organelle and will still contain some contaminating material. Even if you succeed in obtaining, for example, a preparation of mitochondria free of other organelles there will be nuclear DNA adsorbed on the outside of the mitochondria.

Contaminating DNA outside intact organelles can be removed by treatment with DNase, which will not enter intact organelles and so will not degrade their DNA. Unfortunately it can be quite difficult to obtain perfectly intact organelles and some nuclear DNA may be protected within DNase-resistant vesicles. So, for the preparation of completely pure organellar DNA further purification steps are needed.

Also the 'cleared lysate' method for the isolation of plasmid DNA from bacteria, (Section 7.2.2) only enriches the sample for plasmid; it cannot generate completely pure plasmid.

8.2 Approaches to the purification of DNA species

We frequently need to purify our crude preparations of DNA. In this section you will be introduced to some of the methods available for such purifications.

∏   All double-stranded DNA has the same basic structure - the double helix - so what
    properties do you think we might be able to use to distinguish between different
    species of DNA?

If your list included the sequence of nucleotides in the DNA you would be correct, but the sequence does not usually alter the physical properties of the DNA and so cannot be used as the basis of a separation technique.

Different DNA species certainly differ considerably in size. Plasmids are very much shorter than the chromosomal DNA of their host cells and organellar DNA is much shorter than nuclear DNA. You will see in this chapter how size differences allow us to separate different DNAs. When thinking about size you might have considered not just the length of DNA but also its conformation. For example, plasmids normally exist as supercoiled, compact circles of DNA whereas chromosomal DNA, after isolation, is linear. Again, we can exploit such differences for purification of DNA species. Although nucleotide sequence does not affect the physical properties of DNA the base composition does. (Note that in some instances plasmid DNA becomes integrated into the chromosomal DNA. In this case the DNA behaves as though it was chromosomal DNA and this poses problems if we wish to isolate this DNA. Generally we have to treat it as though it was a special collection of host cell genes).

∏ When talking about base composition we refer to 'percentage G/C'. Why do we not usually specify 'percentage G', 'percentage A', 'percentage C' and 'percentage T'?

Remember that G (guanosine) always base-pairs with C (cytosine). Therefore, if 20% of a DNA's bases are G, 20% will be C; so we refer to '40% G/C'. In such a molecule 60% of its bases will be A or T, and we know that each A pairs with a T, so each forms 30% of the bases in the molecule. As you can see, knowledge of the G/C content is sufficient to tell us the percentage of each base in the DNA.

∏ How do you think a molecule with a high G/C content will differ from one with a low G/C value?

From Chapter 6 you will be aware that GC base pairs are held together by three hydrogen bonds but AT pairs have only two such bonds. Consequently the molecule with the higher G/C content will be more stable. This is reflected in the need for a higher temperature to cause separation of the two strands; the DNA is said to have a relatively high melting temperature. Unless the melting temperatures of species to be separated are markedly different, this property cannot usually be exploited for separating species of DNA. However, the buoyant density of DNA increases as its G/C content increases and this can often be used for purification.

melting
temperature

8.3 Purification methods; Separation on the basis of size

We will now consider purification methods which depend on size and density of the molecules.

Molecular biologists spend much of their time separating DNA molecules according to size. Most often they do this analytically, for example, to determine the sizes of DNA fragments produced by a particular restriction enzyme or as a preliminary step in the identification of a piece of DNA containing a specific sequence. Sometimes the separation is carried out preparatively.

The most usual method for separating DNA molecules according to their size is gel electrophoresis. We have already examined this technique in Chapter 1. Since this technique, however, is of vital importance to all molecular biologists we will consider it in more detail here.

8.3.1 Principles of electrophoresis

To carry out electrophoresis, charged species in solution are placed in an electric field. This field causes the species to move; negatively charged ones migrate towards the positive electrode (the anode) and positively charged ones migrate towards the negative 'cathode'.

∏ The speed at which a species moves will depend on several factors. Can you work out what they are likely to be?

The ratio of charge to size will be important. A small molecule with a large negative charge will move faster than a large molecule with the same charge or a small molecule with a lower charge. Another important factor will be the strength of the electric field. this is usually measured as a potential gradient (ie the number of volts per cm). Thus, if the electrodes were placed 20 cm apart, and connected to a power pack delivering 100 V, the potential gradient would be 5 V cm^{-1}. Clearly, the greater the gradient the greater the 'pull' on the charged molecules.

The electrophoresis of DNA is carried out in gels and so the other key factor influencing the rate of migration of molecules is the resistance to movement presented by the gel. In general terms, the smaller a molecule the easier it will be for it to worm its way through the pores of the gel. We can say that, if all molecules have a constant ratio of charge to size, the smallest molecules will move most rapidly through the gel during electrophoresis. This is, in fact, the basis of gel electrophoresis of DNA.

In practice the electrophoresis gel is usually made by pouring a hot solution of molten agarose onto a flat plate with retaining sides (Figure 8.1). As the solution cools the agarose sets to form a gel by hydrogen bonding between its chains. The average pore size of this gel depends on the concentration of agarose; low concentrations produce large pores.

∏ If you wished to separate a mixture of small DNA molecules would you use a high or a low concentration of agarose for your electrophoresis gel?

To separate the molecules you need to ensure that they are all being slowed significantly by the drag exerted by the gel. A gel with very large pores will exert little drag on small DNA molecules and so they will all tend to travel at the same rate through the gel. We therefore use high concentration gels (with small pores) to separate small DNA molecules by electrophoresis.

In practice a typical agarose gel will consist of 0.8% agarose in a 'gel running buffer'. Such a gel is of use with linear DNA molecules of ranging size between about 800 and 10 000 base pairs (bp). For the separation of molecules down to about 200 bp, gels as concentrated as 2.0% are used. At the other extreme, 0.5% gels will allow the separation of molecules up to about 50 000 bp but less concentrated gels are too soft to be used conveniently, thus setting the upper limit for this type of electrophoresis.

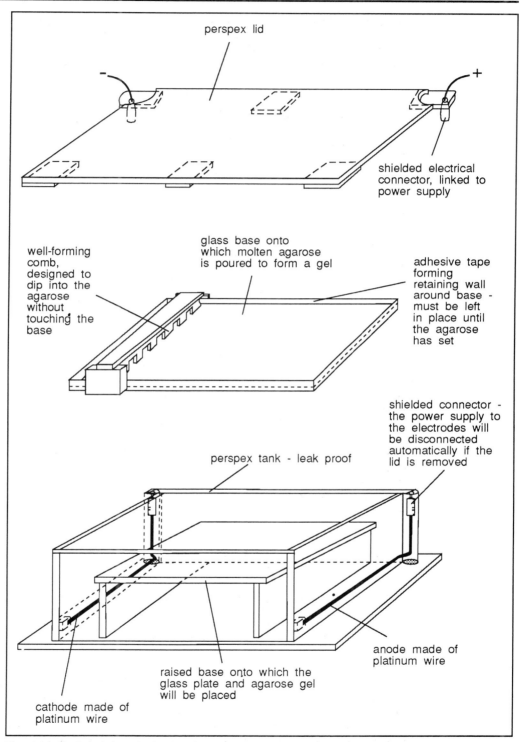

Figure 8.1 Exploded view of submerged gel electrophoresis apparatus. (The apparatus is usually constructed from perspex).

A method of electrophoresis involving regular changes in the direction of the electric field can be used to separate DNA molecules as large as several megabase pairs in size. This technique, known as pulsed field electrophoresis, works because large molecules take longer than small ones to change their direction of migration. The run times tend to be of the order of days, but whole chromosomes of yeast can be separated from each other by pulsed field electrophoresis, making it a very powerful technique.

pulsed field electrophoresis

SAQ 8.1	Select the most appropriate answer(s) to complete the following statement, 'Agarose gel electrophoresis separates DNA molecules according to size because of....'

1) the electric field across the gel;

2) the ratio of charge to mass of the DNA molecule;

3) the pore size of the gel;

4) the attraction between the DNA and the cathode;

5) the pH of the buffer solution.

8.3.2 Electrophoresis in practice

Although it is possible to cast agarose gels between a pair of vertical, glass plates (as for polyacrylamide gels), the horizontal format is easier to prepare and just as effective. Connections between electrode chambers and the gel can be made in a variety of ways, including via wicks of damp filter paper which lie in contact with each end of the gel and dip into the electrode chambers. However, it is far more reliable and easier to use the submerged gel system (Figure 8.1) in which the entire gel is flooded with 'gel running' buffer solution. This provides a reliable electrical contact between electrodes and gel and also helps to cool the gel by convection. Before the gel sets a well-forming comb is placed in position so that its teeth dip into the molten gel near one end but without touching the glass base. When the gel is set the comb can be eased out, leaving wells into which samples can be loaded.

The set gel is placed into the electrophoresis tank with its wells towards the cathode (negative electrode) and is then flooded with a buffer solution such as Tris/borate/EDTA, pH 8.0. Samples of DNA must be mixed with a 'loading mix' which contains Ficoll or sucrose to increase its density so that when we carefully put the DNA solution into the well, it will remain in the well. Bromophenol blue is also included in the 'loading mix' to make the sample easy to see while loading the gel and to act as a marker during the electrophoresis.

∏   Why is the gel positioned with its wells at the cathode end of the tank?

The answer should become obvious if you consider what charge DNA molecules will carry at pH 8.0 (the pH of the electrophoresis buffer solution). The only part of the molecule that will be charged is the phosphate of the 'backbone' of each strand. It is because of the phosphate groups that DNA is an acid and the molecules will consequently carry a negative charge at neutral pH or higher. Such molecules will, of course, be attracted to the positively charged anode. Since all DNA molecules have backbones consisting of alternating phosphate and deoxyribose, all DNA molecules will have the same negative charge per unit length, regardless of nucleotide sequence or length.

∏ For a particular gel, with a fixed voltage across it, what will determine the rates of movement of DNA molecules loaded onto it?

The only difference between molecules, as far as electrophoresis is concerned, will be the size of each molecule. For a mixture of linear DNA molecules we can predict that the distance moved in a given time will increase with decreasing molecular size. This relationship will be discussed more fully in Chapter 9. However, if the mixture contains DNA in several different conformations, such as linear, open circular and supercoiled, it is not possible to predict precisely how they will move. In general we can say that an open circle will move more slowly than a linear molecule of the same length whereas a supercoiled molecule, because it is very compact, tends to move faster than expected from its length.

8.3.3 Visualisation of DNA

After electrophoresis has taken place (gels are often run until the bromophenol blue dye has reached the anode end of the gel), we usually need to examine the gel to discover where each species of DNA has moved. Since the DNA is present at very low levels in the gel and is colourless it can only be seen after staining.

∏ In Section 7.3 you read about a molecule which binds to DNA and is a dye. Can you remember what this molecule is and how it works?

The dye is ethidium bromide. It is a flat molecule that intercalates between the stacked base pairs of DNA and fluoresces in ultraviolet radiation.

Gels soaked in a solution of ethidium bromide (1 mg l⁻¹) for about 10 minutes can be viewed by placing them on an ultraviolet transilluminator. Each species of DNA which has moved by electrophoresis in the gel will be seen as an orange band; each band contains a particular size of molecule (Figure 8.2). It must be stressed that UV radiation is harmful to us, particularly at the intensities produced by transilluminators. Very painful sunburn can result from even brief exposure to UV, eyes can be damaged and skin cancer may occur. It is therefore essential to wear protective masks or use a perspex screen to block the UV radiation and to minimise exposure of hands and arms. Ultraviolet radiation is also harmful to DNA, especially in the presence of ethidium bromide, causing nicks which will degrade the molecules to shorter fragments. For all these reasons you should use the transilluminator for as short a time as possible to view, photograph and, perhaps, cut the gel.

∏ Examine Figure 8.2. Which of the DNA fragments illustrated as bands probably contains the largest fragments of DNA?

You should have concluded that the band nearest the well in tract 2 probably contains the largest fragments because it has migrated least in the electric field. Note however that we have used the word 'probably'. Remember that migration not only depends on the size of the DNA fragments, but also on the conformation of the DNA molecules (open coils, linear or super coils).

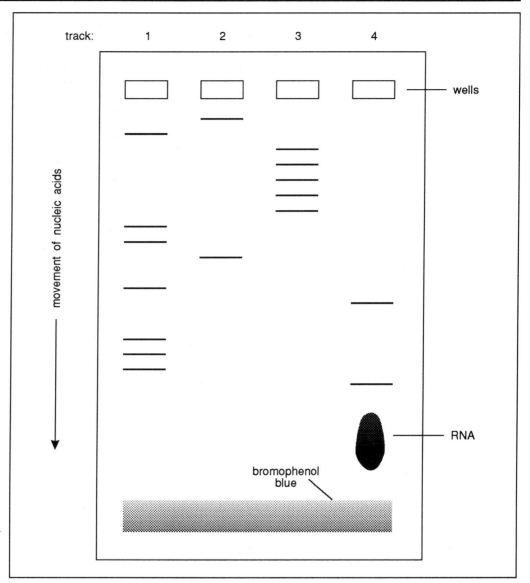

Figure 8.2 Agarose gel stained with ethidium bromide (stylised).Usually photographs of such gels are taken in a transilluminator. The gels appear as dark backgrounds with bright bands which represent the fluorescence associated with the DNA fragments. The figure shown is a commonly adapted stylised representation of such gels.

8.3.4 Recovery of DNA

The analytical uses of agarose gel electrophoresis will be dealt with in Chapter 9. In this chapter we will concentrate on preparative applications.

Once the DNA species have been separated from each other by electrophoresis the next task is to recover the DNA from the gel. There are several ways of doing this. In most cases a slice of gel containing the DNA band of interest is cut out using a scalpel. This must be done as quickly as possible to avoid damaging the DNA by prolonged exposure to UV radiation. Having obtained a piece of gel containing just one band of DNA you can now use electrophoresis once again to move the DNA out of the gel slice.

The way this is done is illustrated in Figure 8.3 from which it should be apparent that the slice is placed in some dialysis tubing with a small volume of gel running buffer solution. The bag is sealed and placed in an electrophoresis tank under more buffer solution. An electric field is applied, causing the DNA to migrate through and out of the gel slice. The DNA then moves rapidly through the buffer solution inside the dialysis bag until it reaches the dialysis membrane. DNA is unable to pass through the minute pores of the membrane and so remains inside the bag. After reversing the direction of the field for a few seconds to pull DNA off the dialysis tubing, the bag is opened and the buffer solution containing DNA is recovered.

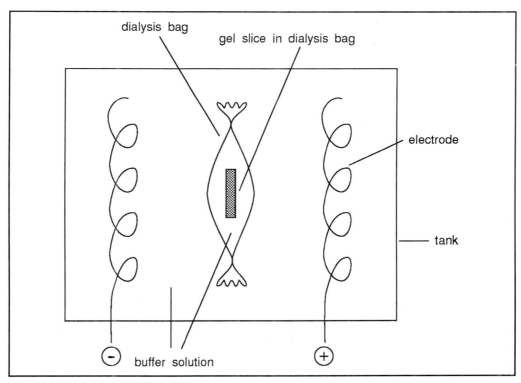

Figure 8.3 Electroelution of DNA

DNA may also be removed from gel slices by treatment of the slice with 'chaotrophic' agents such as concentrated sodium perchlorate or potassium iodide. Such agents disrupt the hydrogen bonds of the agarose gel, causing it to dissolve and allowing DNA to be released into solution. If a special, low-melting point agarose is used for the gel, slices can be melted by heating to about 65 °C. This is not hot enough to melt DNA, allowing recovery of intact, double-stranded DNA from the resulting solution.

The DNA obtained by any of the above methods can be prepared for use by extraction with an equal volume of a phenol/chloroform mixture to remove dissolved agarose, followed by precipitation with ethanol and redissolving in a buffer solution such as TE.

SAQ 8.2	Draw up a list of the items you would need for the recovery of a band of DNA from an agarose gel after electrophoresis, using the technique of electroelution.

8.4 Purification methods; separation on the basis of density

When purifying organelle DNA it is sometimes possible to make use of density differences. Table 8.1 shows the buoyant densities of various types of DNA from a range of plants. It will be seen that, in all cases, there is little difference in density between the nuclear and chloroplast DNAs but there are usually significant differences between these and the mitochondrial DNA densities, reflecting differences in G/C content. Where large differences exist it is possible to separate the organelle DNA from nuclear DNA contaminants by means of density gradient ('isopycnic') ultracentrifugation.

	Buoyant density (g cm^{-3})		
Plant	Nuclear DNA	Chloroplast DNA	Mitochondrial DNA
swiss chard	1.694	1.697	1.705
onion	1.691	1.696	1.706
tobacco	1.695	1.697	1.705
wheat	1.702	1.698	1.707

Table 8.1 Buoyant densities of organelle DNAs isolated from a range of plants.

8.4.1 Principles of isopycnic ultracentrifugation (see also Section 1.2.1)

When a solution of CsCl is subjected to very high gravitational (g)-forces in an ultracentrifuge the molecules of CsCl rapidly redistribute themselves so that there is a higher concentration at the base of the centrifuge tube (farthest from the axis of the rotor) than at its top. Consequently the density of the solution increases down the tube. Any large molecules present in this solution will gradually sink if they are denser than the surrounding CsCl solution or will float 'upwards' if they are less dense. Eventually the molecules may reach a point in the tube where the density of the CsCl solution is exactly the same as their own density. At this point there will be no tendency to sink or float and so the molecules will come to rest. Molecules with different densities will therefore end up in bands at different positions along the tube. In a conventional, swing-out rotor it may take as long as two days for all molecules to reach their resting point.

The bands of DNA within the centrifuge tube will not be visible so it is necessary to pump out the contents gently and monitor absorbance at 260 nm to identify fractions containing DNA.

8.4.2 Isolation of supercoiled DNA

We frequently need to isolate plasmid from bacteria. The cleared lysate method for obtaining a sample enriched for plasmid has already been described (Section 7.2.2) but such samples still contain some chromosomal DNA.

Unfortunately the buoyant density of a plasmid is normally very similar to that of the chromosomal DNA and so the method described above (Section 8.4.1) would not be able to resolve the two species of DNA. However, it is easy to produce a difference in density between plasmid and chromosomal DNA. To understand how this is done it is important to remember that the native form of plasmids consists of highly compact, supercoiled circles of DNA. To understand such supercoils try the following experiment with a piece of string: hold one end of the string or tie it to a firm support, then start twisting the other end. As you do this you will notice that considerable tension builds up in the string. If the ends of the string are allowed to come closer together the string will suddenly appear to tie itself in knots, forming a relatively compact tangle of string.

supercoils

∏ What happens when you release one end of the string? What happens if you stick, clamp or tie the free ends together?

You should find that releasing one end of the string allows it to release tension by unwinding. By contrast, a string which has no free ends, because they are fixed together, is unable to unwind and so remains in its compact, 'supercoiled' state.

Plasmids become supercoiled by a similar process. The circular DNA double helix is unwound to some extent, creating tension which drives the formation of compact 'tangles' of DNA. Because the plasmid has no free ends, being a covalently closed circle (except during unwinding), the supercoiling remains in place.

∏ What happens to DNA when ethidium bromide binds to it? Will this binding be equally easy with linear and supercoiled DNA?

You should have remembered that ethidium bromide causes some unwinding of DNA as it becomes intercalated between stacked base pairs (Section 7.3). Linear DNA has free ends and so can unwind easily, allowing the binding of large quantities of dye per unit length. Supercoiled plasmid is already under tension owing to its extra negative turns (ie it has been unwound to some extent) so further unwinding will be comparatively difficult. Consequently, even in saturating levels of ethidium bromide, less dye can bind to a given length of supercoiled plasmid than to the same length of linear or open circle DNA.

Open circles are circular molecules of DNA in which there is no supercoiling, most usually because there is at least one point in at least one strand of the DNA where the backbone is broken('nicked'), allowing free unwinding of supercoiled molecules.

In addition to acting as a dye for the visualisation of DNA, ethidium bromide causes a decrease in the density of DNA to which it binds. The more that it is bound, the greater the decrease in density.

∏ Rank the following in order of how much ethidium bromide will be bound per unit length:

• fragments of chromosomal DNA;

• open circle plasmid;

• supercoiled plasmid;

• linear plasmid.

You have seen how supercoiled plasmid is limited in its possible binding of dye, so this must be the lowest in the ranking. The remaining types of DNA have something in common - they all have 'free ends', even if only as a result of a single nick in one strand; hence all can unwind freely as ethidium bromide binds and all are joint first in the ranking!

The important consequence of all this is that, in the presence of excess ethidium bromide, supercoiled plasmid will remain relatively dense but all other forms of plasmid and all pieces of chromosomal DNA will be significantly reduced in density. If we now carry out isopycnic ultracentrifugation of DNA from a cleared lysate in the presence of excess ethidium bromide, supercoiled plasmid will band beneath all other types of DNA. The DNA bands are easily seen when viewed by UV radiation, owing to the fluorescence of the bound dye.

isopycnic ultra-
centrifugation

The lower, supercoiled plasmid band (Figure 8.4) can then be recovered, free of contaminating DNA, by drawing if off through a syringe needle, either from above or by side puncturing of the centrifuge tube just beneath the plasmid band. Ethidium bromide is readily removed from the DNA by partitioning against iso-amyl alcohol. After several cycles of extraction all dye will have been removed from the plasmid.

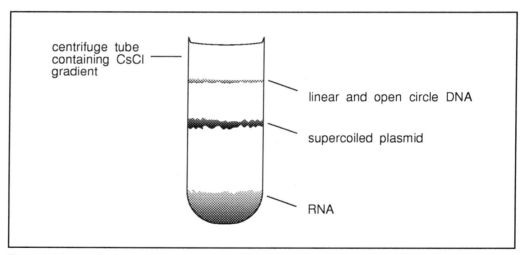

Figure 8.4 Isopycnic ultracentrifugation of plasmid preparation. The preparation was already enriched for plasmid and free of protein. After adding CsCl and ethidium bromide it was centrifuged at 38 000 rpm in a swing-out rotor for 40h. UV radiation revealed a thick band containing pure, supercoiled plasmid and a less dense, less intense band containing low levels of linear and open circle DNA species. At the base of the tube was a diffuse glow from dye bound to RNA.

CsCl can be removed by dialysis of the solution or by dilution with buffer solution followed by precipitation with two volumes of ethanol. Dilution is needed to avoid precipitation of CsCl by the ethanol. Another way to remove CsCl is by the use of desalting columns. These are miniature gel exclusion columns packed with Sephadex G-10 or equivalent and equilibrated with buffer solution. When the DNA sample is forced through the column with a syringe or by centrifugation, the DNA, which is too large to enter the gel matrix, emerges in buffer solution leaving the CsCl in the column. It is, of course, important that the volume of DNA solution forced through the column is slightly less than that of the column; a larger volume would result in some CsCl emerging at the tail end of the DNA fraction.

SAQ 8.3

The following samples were subjected to isopycnic ultracentrifugation in the presence or absence of saturating concentrations of ethidium bromide:

1) pBR322 plasmid DNA cut into two pieces, one twice the length of the other;

2) pBR322 plasmid DNA in its supercoiled and open circular forms (equal concentrations);

3) plasmid DNA, buoyant density 1.685 g cm^{-3}, in its supercoiled and open circular forms, with an equal concentration of linear, chromosomal DNA, buoyant density 1.684 g cm^{-3}.

For each sample work out how many bands of DNA would be produced by the ultracentrifugation, with and without ethidium bromide, and list the contents of each band, starting with the band nearest the base of the centrifuge tube.

Summary and objectives

DNA molecules can be separated according to size by preparative gel electrophoresis. DNA is recovered from the gel by electroelution, chaotropic reagents or use of low melting point agarose. Isopycnic ultracentrifugation is able to separate DNA molecules according to buoyant density, and hence base composition, provided the molecules have significantly different densities. Supercoiled molecules can be separated from other DNA by ultracentrifugation in the presence of ethidium bromide.

On completion of this chapter you should now be able to:

* give an account of the principles of electrophoresis of DNA;

* describe and understand how to recover DNA from a gel after gel electrophoresis;

* explain the principles of isopycnic ultracentrifugation, including an appreciation of when it is appropriate to use ethidium bromide;

* identify different bands of DNA separated by isopycnic centrifugation.

Chemical composition, structure and properties of DNA

Chemical composition, structure and properties of DNA

9.1 Introduction

In this chapter the most commonly used methods for determination of the length, conformation, base composition, complexity and nucleotide sequence of a DNA sample are described and their theoretical basis is explained.

The general structure of DNA is well established and does not vary from one source to another, although the exact shape of the molecule is now known to be influenced by the nucleotide sequence. The elucidation by Watson and Crick of the double helix structure of double-stranded DNA is one of the landmarks of science and, although some significant variants of their model have been described, the modern molecular biologist does not need to redetermine the secondary structure of DNA. An outline of DNA structure is given in Chapter 6.

The variables that you are most likely to need to analyse are:

- the length of the DNA;
- its conformation (linear, circular, supercoiled);
- its base composition;
- its complexity;
- the sequence of its nucleotides.

9.2 Measurement of DNA length

∏ A method for the separation of DNA molecules on the basis of size has already been described in Section 8.3.2. Can you remember what this method is?

Electrophoresis of DNA in agarose gels will separate DNA molecules according to size. Remember that the DNA molecules experience resistance to their movement as they pass through the pores of the gel; hence a high concentration gel, with small size, will retard DNA more than a low concentration gel.

agarose gel electrophoresis

The preparative uses of agarose gel electrophoresis have been described. However, the method can also be used for very accurate estimation of DNA lengths.

∏ Figure 9.1 shows the positions of DNA fragments of known lengths in an agarose gel after electrophoresis. Using only track 1 of the gel, plot a graph of distance from the loading well for each band of DNA against the fragment length in kilobase pairs (thousands of base pairs, or kb). Then plot each distance against the logarithm of the fragment length. Compare your graphs with those in Figure 9.2. Which plot gives the more linear relationship?

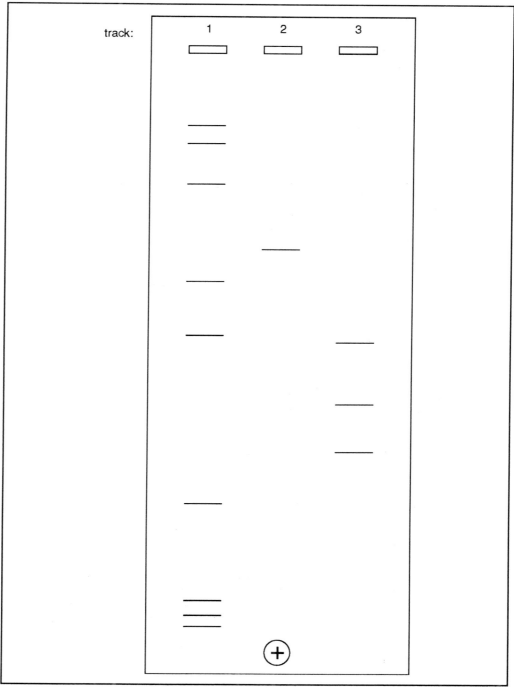

Figure 9.1 Gel electrophoresis of DNA samples. Track 1 contains a mixture of restriction fragments of the following sizes (kb): 1.41, 1.98, 2.26, 3.16, 5.62, 6.75, 9.46, 12.9, 23.7. Track 2 contains a single species of linear DNA. Track 3 contains only the plasmid pBR322.

Over a useful range of DNA lengths is it found that a plot of distance moved against the logarithm of DNA length gives an approximately linear relationship (Figure 9.2). It is hoped that you remembered that the larger sized molecules moved a shorter distance!

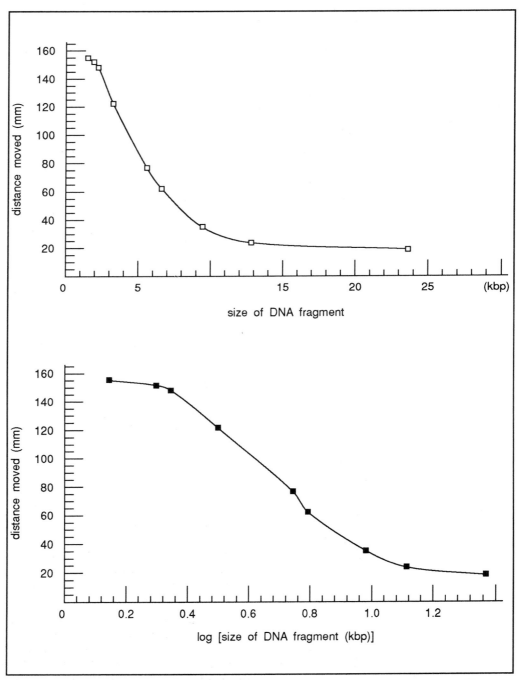

Figure 9.2 Plots of distance moved against fragment size —□— , against log (fragment size) —■— .

Fragments of known length can readily be obtained by digesting a well characterised DNA, such as that from the bacteriophage λ, with one or more restriction enzymes. The example in Track 1 of Figure 9.1 was obtained by digesting λ DNA with a mixture of restriction enzymes. Such preparations can then be used as markers for calibration of a gel in which 'unknown' samples are run in tracks alongside markers.

∏ Using the calibration curve you plotted for Figure 9.1, estimate the length of the DNA that has been run in track 2.

You should have obtained a value of about 7.5-7.6 kb. If your result was significantly different from this, follow these steps:

• measure the distance between the DNA band and the well in Track 2;

• find this distance on your calibration graph and, from the curve, read off the corresponding log of the DNA length. Take the anti-log of this figure. This should give you the size of the DNA in kb.

By careful measurement using large photographs of large gels it is possible to obtain very accurate estimates of the sizes of DNA molecules. Such estimates are frequently needed by the molecular biologist. For example, they are essential for the accurate mapping of restriction sites on a length of DNA (BIOTOL text, 'Techniques for Engineering Genes').

∏ Track 3 of Figure 9.1 contains only the plasmid pBR322. Can you suggest why three bands of DNA are visible?

To understand this result you must remember that agarose gel electrophoresis separates DNA according to size. If all the molecules are linear then separation will be according to length.

However, plasmids can adopt several conformations - linear, open circular or supercoiled - all of which will have the same length but different sizes. You may recall that supercoiled molecules are very compact (Section 8.4.2) and so move faster through the gel than expected from their length; open circles are retarded more than linear forms of the same length.

SAQ 9.1   Estimate the length of pBR322 from Figure 9.1 and your calibration curve.

You should have noticed on your calibration curve that the relationship between log (size) and distance migrated by DNA molecules is not linear at extremes of size. The range covered by the linear portion of the plot depends on the concentration of the gel: for accurate size estimation of small DNA molecules a high concentration gel (up to about 2%, w/v) should be used whereas concentrations as low as about 0.5% will be more appropriate for large molecules. Molecules larger than about 50 kb must be sized by use of pulsed field electrophoresis (Section 8.3.1).

When the DNA molecules fall below about 200 bp (base pairs) in size even the most concentrated agarose gels will fail to resolve them. In such cases we must make use of polyacrylamide gels (Section 1.3.4) since these have a much smaller pore size than can be achieved with agarose. Such gels are used routinely for checking the sizes and purity of oligonucleotides, for sizing small fragments of DNA produced by restriction enzymes and, as you will see later in this chapter, for DNA sequencing.

9.3 Analysis of DNA conformation

It is relatively easy to determine if a species of DNA is circular or linear. Circular molecules will be resistant to digestion by exonuclease III (from *E. coli*) since this enzyme normally only attacks free 3'-termini. Thus the DNA under test can be incubated with exonuclease III and then analysed by gel electrophoresis to see if its size has been reduced.

A similar approach can be taken to distinguish between single-stranded and double-stranded DNA. In this case we can make use of the enzyme S1 nuclease (from *Aspergillus oryzae*) which specifically hydrolyses single-stranded DNA to mononucleotides.

topoisomerase   A class of enzyme called topoisomerase can be used to determine if a circular molecule is supercoiled or relaxed. These enzymes temporarily break one strand of a supercoiled molecule, allowing the DNA to relax in a controlled manner, thereby releasing some of the tension of supercoiling; the DNA remains circular throughout this process. Relaxed molecules will not be affected by treatment with topoisomerase but supercoiled molecules will be converted into a mixture of circular molecules with varying degrees of supercoiling. These will migrate more slowly than the original, fully supercoiled molecule during electrophoresis giving rise to a 'ladder' of bands on the gel.

SAQ 9.2

Two samples of DNA, labelled A and B were analysed by agarose gel electrophoresis before and after treatment with a selection of enzymes. Untreated sample A generated two bands of DNA and B gave only one band. After extended treatment with S1 nuclease sample A gave only one of the original bands; sample B was unchanged. After exonuclease treatment sample A gave no bands, only a low molecular mass smear, sample B was unchanged.

Interpret these results and suggest how you could investigate sample B further.

9.4 Analysis of base composition

9.4.1 Analysis of hydrolysis products

It is possible to measure base composition by direct analysis. Unlike RNA, DNA is not hydrolysed by alkali but it is susceptible to attack by acids. Dilute acid will cause removal of purine bases from the DNA ('depurination') and the resulting molecules will tend to be hydrolysed by a reaction involving the C-l' sugar of the deoxyribose wherever a purine is missing. Such hydrolysis, if carried out for a short time using dilute acid, can be a useful way to break high molecular mass DNA into shorter fragments and is often included prior to Southern Blotting of an agarose gel in which DNA must be transferred efficiently out of the gel onto a membrane, to which it will bind in a pattern identical to that in the gel. Unless large DNA molecules are fragmented they will not move out of the gel rapidly enough for complete transfer to the membrane.

More extensive acid treatment will release all the bases from a DNA sample and these can then be separated from each other by a variety of chromatographic methods, including HPLC, and quantified to give a precise base composition.

9.4.2 Denaturation of DNA

Complementary strands of DNA base-pair to form double-stranded molecules. You should remember that the bases are held together by weak hydrogen bonds and it is only because there are so many of these bonds between the two strands of a typical DNA molecule that the structure is so stable at normal temperatures.

∏ If a sample of double-stranded DNA is warmed up, what do you think will happen to it?

At first very little change will occur. Even though, at any instant, there may be a localised separation of short lengths of the two strands caused by the increased kinetic energy of the molecules, the vast majority of each DNA molecule will remain double-stranded. However, as the temperature of the sample rises a temperature will be reached at which the locally denatured regions start to coalesce and cause extensive strand separation; a slight further rise in temperature will result in complete denaturation of the DNA to its single-stranded form.

This denaturation can be monitored by measuring the absorbance of the sample at 260 nm. A given solution of DNA will have an A_{260} when fully denatured, that is about 30% higher than its value when entirely double-stranded. This increase in absorbance caused by denaturation is known as the hyperchromic effect. A typical plot of A_{260} against temperature is show in Figure 9.3. The rapid rise in absorbance for a small temperature increment is reminiscent of the sudden melting of a pure crystalline compound; hence the denaturation profiles of DNA are often referred to as 'melting curves' and the temperature at which 50% denaturation occurs is called the 'melting temperature' or T_{m} of the DNA sample.

hyperchromic effect

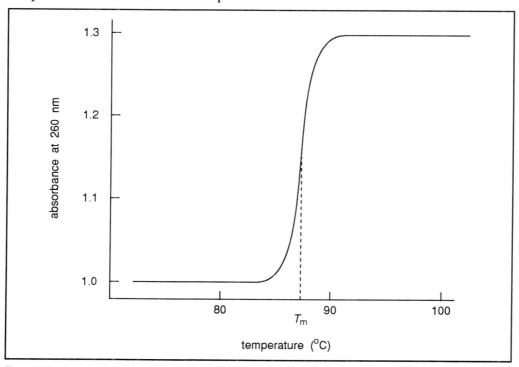

Figure 9.3 Melting curve of a pure DNA sample.

⊓ Different species of DNA exhibit different T_m values. Can you suggest what might
    be responsible for these differences?

The stability of a double-stranded DNA will depend on the number of hydrogen bonds per unit length holding the chains together. The greater the number of bonds the higher will be the T_m. Since adenine-thymine base pairs are linked by only two hydrogen bonds but cytosine-guanine interactions involve three such bonds it should be obvious that a species of DNA with a high G/C content will also have a high T_m.

In fact, provided standard ionic conditions are used, there is a fixed relationship between T_m and base composition that allows melting temperature determinations to be used as an indirect means of determining the base composition of DNA. This relationship is:

$$T_m = (39.7 \times GC) + 69.0 \qquad\qquad (E - 9.1)$$

where T_m is in °C and GC is the content of guanosine plus cytosine as a fraction of all bases in the DNA.

| SAQ 9.3 | Calculate the percentage of adenosine bases in a DNA sample whose T_m was found to be 85.4°C. What have you assumed in your calculation?

9.5 Analysis of DNA complexity

Once a DNA sample has been denatured it can be cooled and allowed to renature. Rapid cooling of the DNA will tend to produce fairly random pairing of short regions of imperfectly matched strands. As long as a few bases every now and again along a strand can pair with complementary regions in another strand a stable structure will result. There will, of course be considerable lengths of single-stranded DNA looping out between the paired regions and the net result will usually be a very high molecular mass tangle of molecules. The reason for this non-specific pairing is that insufficient time is allowed for truly complementary strands of DNA to collide in the correct orientation for accurate base-pairing and incorrect pairings are relatively stable at the low temperature used for renaturation.

If renaturation is allowed to occur at a temperature only about 10°C below the T_m of the DNA, double-stranded structures formed by incorrect base-pairing will not be stable and so will rapidly separate again. Only perfectly base-paired DNA will be sufficiently stable to survive under these stringent conditions.

stringent conditions

⊓ If you monitored the renaturation of λ DNA (49 kb in size) and, in a separate tube,
    human DNA (2.3×10^6 kb), under conditions where the concentrations of DNA
    (mass per unit volume) were identical in the two experiments, which do you think
    would renature more rapidly?

Since there is the same mass of DNA per unit volume (cm^3) there will be a much higher number of the small λ molecules in each cm^3 than there will be of the huge human genome. Thus the chance of two complementary strands of DNA colliding in the correct orientation for accurate base-pairing will be vastly greater for λ than for human DNA.

The argument remains the same even if, as is usual for such experiments, the DNA molecules are broken or cut randomly into pieces a few kb in size.

By measuring the rate of renaturation of a DNA sample it is, therefore, possible to estimate the degree of complexity of the sample. For the total DNA from an organism the complexity can be regarded as equivalent to the size of that organism's genome. It is normal to plot the degree of renaturation against time multiplied by the nucleotide concentration, or $C_o t$. Such curves are known as 'Cot' curves and the value of $C_o t$ at which the sample is 50% renatured is known as $(C_o t)_{0.5}$ or 'Cot a half'; it is this value which is taken as the measure of the sample's complexity.

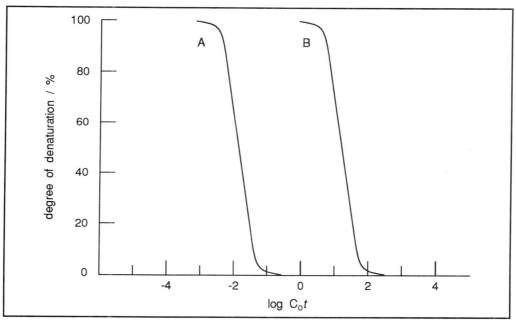

Figure 9.4 Renaturation kinetics of two DNA samples. Samples from organisms A and B were denatured and their rates of renaturation were monitored at a temperature 10°C below their T_m values. The degree of renaturation was plotted against log $C_o t$ where C_o is the concentration of the nucleotides (mol l^{-1}) and t is the time (s).

SAQ 9.4

Two Cot curves are presented in Figure 9.4, produced by analysis of the DNA from two different organisms. Which organism has the larger genome?

9.6 Sequencing of DNA

The availability of rapid and accurate methods for determining the sequence of nucleotides in a DNA sample has transformed our understanding of the way in which genes are organised and expressed. Although there are several methods for sequencing DNA the method devised by Sanger, known as the 'dideoxy' or 'chain termination' reaction is now the most widely used.

9.6.1 The Sanger or dideoxy method

In Sanger's method the DNA to be sequenced is commonly cloned in a DNA vector derived from the bacteriophage M13. The point of this step is that large amounts of single-stranded DNA can be recovered very easily from the medium surrounding bacteria infected with the M13. Most importantly, all the molecules are identical; they are not a mixture of two complementary forms of the molecule but are all the 'plus' strand.

The M13 DNA used for sequencing experiments has been modified so that it includes a special 'multiple insertion site' at which the DNA to be sequenced can be inserted. The M13 DNA on either side of this site has been sequenced and oligonucleotides have been synthesised that are complementary to short stretches of 'plus' strand DNA on the 3' side of the insertion site. Such an oligonucleotide, if incubated with the single-stranded M13 (containing its DNA insert) under stringent conditions of hybridisation (ie at a temperature and ionic strength at which only exactly complementary strands will form stable, double-stranded structures) will hybridise specifically to its complementary region on the M13 DNA.

The oligonucleotide can now act as a primer for the synthesis of DNA by the enzyme DNA polymerase. This enzyme will only add nucleotides to a pre-existing DNA (or RNA) strand, hence the need for a primer. Strand synthesis proceeds in a 5' to 3' direction and the new strand is complementary to the M13 template. It should be evident from Figure 11.5 that, since the primer binds close to the 3' end of the multiple insertion site, one of the first pieces of DNA to act as a template will be the DNA inserted in that site. Thus an early product will be DNA complementary to the fragment to be sequenced.

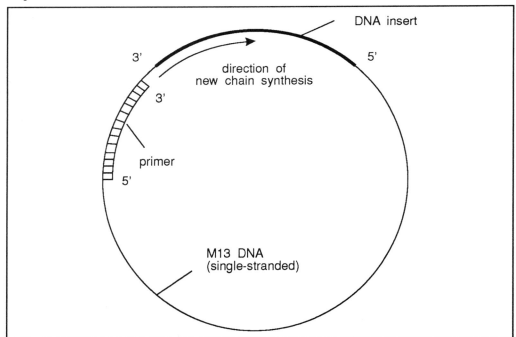

Figure 9.5 Hybridisation of primer to M13. The single stranded M13 molecule contains DNA to be sequenced inserted within its 'multiple insertion site'. To the 3' side of this insert is a sequence complementary to the primer. Once the primer has annealed to the M13, new chain synthesis can take place from the 3' end of the primer, using the M13 and its insert as template.

If the above reaction is carried out using deoxyribonucleotide precursors that are radiolabelled the newly synthesised DNA will also be radiolabelled. For sequencing, the DNA synthesis is carried out in four different tubes, each containing polymerase and all four deoxyribonucleotides, at least one of which is radiolabelled. In one tube there is also some dideoxyadenosine triphosphate, in the next there is some dideoxycytidine triphosphate, with dideoxyguanosine triphosphate in the third and dideoxythymidine triphosphate in the last tube. The significance of the dideoxyribonucleotides is that they lack a 3' hydroxyl group. When a normal deoxyribonucleotide is added to a growing DNA strand its 3' hydroxyl forms the point of attachment of the next nucleotide, via a 3',5'-phosphodiester linkage. If a dideoxyribonucleotide is incorporated into a growing DNA chain there will be no point of attachment for the next nucleotide and so growth of the chain will be terminated at that point.

Since each of the sequencing reaction tubes contains a mixture of a dideoxyribonucleotide and its corresponding deoxyribonucleotide there will be a competition for incorporation into the growing DNA chains. The chances of chain termination will therefore depend on the relative levels of the two types of nucleotide. Consequently the newly synthesised DNA in the first tube will consist of a mixture of molecules of various lengths, all terminating with a dideoxyadenosine; other tubes will contain DNA terminated with their corresponding dideoxynucleotide.

∏ Write down all the possible radioactive products that could be formed in the tube containing polymerase, deoxyribonucleotides (dCTP radiolabelled) and dideoxyadenosine triphosphate and the following DNA template and its associated primer:

3'-CGGTATTACCGAATTCGGGCATTTCAAGGACTA (template)

5'-**GCCATAAT** (primer)

All the products will begin with the primer and will then extend to a dideoxyadenosine. Since the dideoxy- and deoxy- forms of the nucleotide are both present the primer will be extended step-by-step as follows:

5'**GCCATAAT**G

5'**GCCATAAT**GG

5'-**GCCATAAT**GGC

5'-**GCCATAAT**GGCT

5'-**GCCATAAT**GGCTT

5'-**GCCATAAT**GGCTTA.

Since dideoxyadenosine triphosphate is present in the mixture there is a chance that some molecules will have received the dideoxy form at this point and so growth of the chain will be terminated. Thus the first 'product' in this reaction tube will be:

3'CGGTATTACCGAATTCGGGCATTTCAAGGACTA

5'-**GCCATAAT**GGCTTA.

Most DNA molecules will, however, have been extended by the addition of a normal, deoxyadenosine, and so will continue to increase in length. Wherever adenosine is inserted into the growing chain there will be a chance of chain termination. Consequently, if we ignore the original template, we can write down a set of potential products as follows:

5'-**GCCATAAT**GGCTTA

5'-**GCCATAAT**GGCTTAA

5'-**GCCATAAT**GGCTTAAGCCCGTA

5'-**GCCATAAT**GGCTTAAGCCCGTAA

5'-**GCCATAAT**GGCTTAAGCCCGTAAA

5'-**GCCATAAT**GGCTTAAGCCCGTAAAGTTCCTGA

This exercise could be repeated for each of the other tubes, giving four sets of products, each containing newly synthesised, radioactive molecules terminating with a given dideoxyribonucleotide.

The four samples synthesised as above are then denatured and subjected to electrophoresis under denaturing conditions (high temperature and high concentration of urea) on adjacent tracks of a very high resolution polyacrylamide gel. After electrophoresis the positions of radioactive DNA bands are revealed by autoradiography of the gel. Differences in DNA length of only one nucleotide are resolved by the long, thin gels used for sequencing.

Π If the products you wrote down in the previous exercise were separated on such a gel, what would the autoradiograph of this gel look like? To simplify your task, assume that an increase in length of one nucleotide results in a fixed decrease in distance moved, regardless of the total length of the molecule.

The lengths of the radioactive chains of DNA, including the primer, are: 14,15,22,23,24 and 32 nucleotides. Thus the spacing of radioactive bands will be as shown below.

```
markers:          | | | | | | | | | | | | | | | | | | | | | |
chain lengths:    15          20          25          30          35
bands:            | |                   | | |                   |
```

If we had to deduce the DNA sequence from this autoradiograph we could write the following:

A,A,6 unknowns, A,A,A,7 unknowns, A.

To fill in the 'unknowns' we need the patterns of bands produced by the other three samples. Note that we start reading the sequence from the smallest fragment (ie the band that has moved farthest through the gel) and that the sequence we produce is that of the newly synthesised chain, not of the template.

SAQ 9.5

Figure 9.6 shows a pattern of bands revealed by autoradiography of a sequencing gel. Write down, as far as possible, the sequence indicated by this autoradiograph.

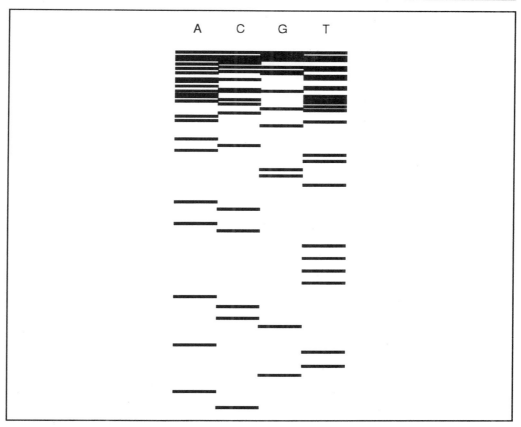

Figure 9.6 Autoradiograph of a sequencing gel. The tracks are labelled according to the dideoxyribonucleotide that was present in the corresponding reaction mixture. Thus A indicates the sample from the reaction containing dideoxyribonucleotide.

9.6.2 The Maxam and Gilbert method

An alternative to the dideoxy method of DNA sequencing is that of Maxam and Gilbert. We described this method for determining the nucleotide sequence of RNA. Here we will remind you of the principles. This method involves end-labelling of the DNA to be sequenced, followed by treatment in four separate tubes with chemicals that will cleave the DNA preferentially at guanine, guanine and adenine, thymine and cytosine, or cytosine. These cleavage reactions are carried out under conditions in which very few sites are hydrolysed in a given molecule; thus each of the reaction tubes contains a mixture of radioactive products, all ending at positions originally occupied by the same type of nucleotide.

Let us examine the details of this procedure.

First the DNA restriction fragment is radioactively labelled at one end with ^{32}P-phosphate. This can for example be achieved at the 5' ends by treatment with alkaline phosphatase to dephosphorylate the DNA. This is followed by incubation with 5'-(-γ-^{32}P) ATP and polynucleotide kinase. This results in the 5' ends becoming labelled with ^{32}P-phosphate. The labelled DNA sample is divided into aliquots and aliquots are treated with a variety of chemicals which modify the bases in the DNA. Table 9.1 gives some examples of the chemical reagents which are used to alter particular bases in

DNA. A good example is dimethyl sulphate which reacts with the nitrogen at position 7 in guanine. This reagent methylates guanine. After chemical modification the altered base is then removed from the sugar phosphate backbone of the DNA. The strand is then cleaved with piperidine. Piperidine hydrolyses the strand at any sugar residue lacking the base. This process is illustrated in Figure 9.7 showing how dimethylsulphate is used to produce specific fragments of DNA.

Base(s) modified	Reaction	Process of altered base removal	Strand cleavage
G	dimethylsulphate	piperidine	piperidine
C	hydrazine + NaCl	piperidine	piperidine
G+A	acid	depurination by acid catalysis	piperidine
T+C	hydrazine	piperidine	piperidine
C<A	NaOH	piperidine	piperidine

Table 9.1 Reactions used in the Maxam and Gilbert procedure for DNA sequencing.

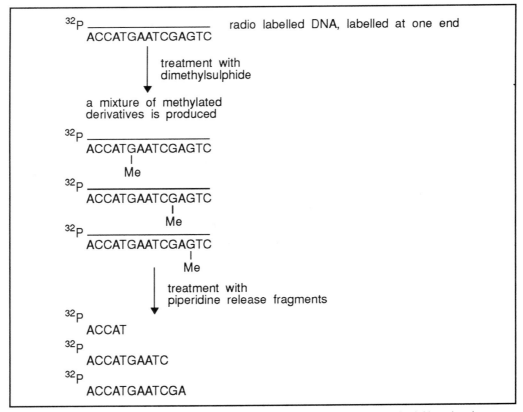

Figure 9.7 Sequence for the degradation of DNA using the Maxam and Gilbert method. Note that the sequence shows the events for chemical cleavage of the DNA at G residues. Notice only the terminal fragments are labelled with ^{32}P.

In Table 9.2 we give a few more details of the chemical modification used in the Maxam and Gilbert procedure.

Base	Modification
G	N7 is methylated by dimethyl sulphate at pH 8.0. This makes the C8-C9 very susceptible to hydrolysis by base.
C	Hydrazine opens the cytosine pyrimidine ring which subsequently recycles into a five membered ring. The five membered ring is readily removed by hydrolysis. The presence of 1.5 mol l^{-1} NaCl prevents the ring opening in thymine.
A+G	At pH 2.0, piperidine formate protonates the nitrogen atoms in the purine rings and this weakens the glycosidic bonds. Adenine and guanine are subsequently easily removed.
C+T	Hydrazine opens the pyrimide rings (see C).
C>A	1.2 mol l^{-1} NaOH at 90°C causes substantial cleavage at A, but only a limited amount of cleavage at C.
Sugar phosphate chain	The sugar phosphate chain is hydrolysed at sites where bases have been modified by 1 mol l^{-1} piperidine at 90°C.

Table 9.2 Chemical modifications used in the Maxam and Gilbert procedure.

Separation of the products on a sequencing gel and subsequent autoradiography are exactly as described for the dideoxy method. Interpretation of the autoradiograph is similar to that already described, but a band may be seen at the same position in more than one track. For example, a band in both the 'G' and 'G & A' tracks indicates a 'G'; a band in the 'G & A' but not the 'G' track indicates an 'A'.

The main attraction of Maxam and Gilbert sequencing is that it does not involve the enzymatic synthesis of a new DNA strand and so is not affected by some of the 'difficult' sequences which can cause artifacts in the dideoxy method. However, since the DNA is only end-labelled it can never be as radioactive as the uniformly labelled products of the dideoxy method; thus more DNA and longer exposure times are needed for Maxam and Gilbert sequencing.

Summary and objectives

Gel electrophoresis can be used to determine the size of DNA molecules. The conformation of DNA will affect its mobility in an electrophoresis gel so linear calibration standards can only be used with linear 'unknown' DNA molecules. Enzymes can be used to determine the conformation of DNA molecules.

Acid hydrolysis of DNA liberates bases which can be estimated chromatographically. Denaturation of DNA by heat allows determination of melting temperatures; these can be related to base composition. The rate of renaturation of a DNA sample, as determined by Cot curves, allows us to estimate the complexity of the DNA.

The dideoxy method of DNA sequencing is used widely and requires little DNA. Maxam and Gilbert sequencing is of particular use when 'difficult' base sequences cause artifacts with the dideoxy method. Sequences are read from autoradiographs of the sequencing gel.

On completion of this chapter you should now be able to:

- determine the length of a DNA molecule from its behaviour during agarose gel electrophoresis;

- choose appropriate methods and interpret their results for the determination of the conformation of DNA samples;

- describe and understand the principles of methods for determining the base composition of a DNA sample;

- explain how Cot curves can be used to determine the complexity of a DNA sample;

- describe the principles of the Sanger and Maxam & Gilbert methods for DNA sequencing;

- read a DNA sequence from an autoradiograph of a sequencing gel.

Extraction and estimation of ribonucleic acid (RNA)

Extraction and estimation of ribonucleic acid (RNA)

10.1 Introduction

Ribonucleic acids play important roles in the process of protein synthesis. They are seen as the molecules which enable the nucleotide sequence of DNA to be converted into the sequences of amino acids found in proteins. They are of widespread occurrence in nature.

In this chapter we shall consider the extraction and purification of RNA and its estimation. More details of the structure and occurrence of nucleic acids are covered in the BIOTOL text, 'The Molecular Fabric of Cells', and their function in protein synthesis is described in the BIOTOL text, 'The Infrastructure and Activities of Cells'.

10.2 The extraction of RNA

10.2.1 Introduction

What do we require of an RNA extraction procedure?

Firstly, that the RNA is pure (or, perhaps more practically, pure enough) and secondly, intact. The first point is fairly obvious but the latter possibly less so. Let us take these points in turn.

∏   Without knowing anything about RNA extraction procedures, what do you think will be the most likely contaminant in a crude RNA preparation?

The answer is DNA. It has so many similar properties: charge, sugar-phosphate polymer, nitrogenous bases and high molecular mass. It will, therefore, co-extract with RNA. Of course, other contaminants are likely, as follows:

* proteins; particularly basic proteins as they will be positively charged and therefore have strong ionic interaction with the negatively charged phosphate groups;

* carbohydrates; these are very varied and depend on cell type. The most likely problems are going to come from high molecular mass carbohydrates such as those used for energy stores in animal cells (ie glycogen) and cell wall carbohydrates found in micro-organisms and plants;

* lipids; theoretically they should be less of a problem, why? Lipids are either wholly, or at least predominantly, non-polar (ie hydrophobic). Therefore, it should be relatively simple to devise a two-phase solvent extraction system that can exploit this fundamental difference from nucleic acids;

* small molecular mass compounds.

∏   There is a wide variety of small molecules found in cellular extracts. What do they all have in common that is different from nucleic acids?

They are low molecular mass, so it should be possible to devise a procedure based on solubility or chromatography to separate them from RNA and DNA.

∏ Having said that, which low molecular mass compounds are likely to present the most problem?

The nucleosides and nucleotides (ATP, UDP, cAMP, etc) are the precursors and products of RNA synthesis and degradation and are likely to co-purify with DNA and RNA. Size is the only property distinguishing nucleotides from the macromolecules required.

∏ The second objective stated above was to extract intact molecules. What do you think will be the most likely cause of RNA breakdown?

Enzymic degradation, namely ribonuclease (RNase) activity. Ribonucleases come from a number of sources: the tissue or cells being extracted and contamination from perspiration on fingers or dirty glassware.

precautions in handling RNA
The following precautions are necessary:

- experiments with RNA are carried out with the utmost cleanliness and attention to detail; wearing disposable gloves is a precaution against finger nucleases;

- glassware, plasticware and solutions should be autoclaved where possible or;

- rinsed with 0.5% diethylpyrocarbonate (DEPC, a powerful protein denaturant) followed by incubation at 100°C for five minutes (alternatively glassware can be heated to 150°C for 30 minutes without DEPC treatment);

- RNA is stored under conditions that minimise RNase activity, eg low temperature, in the presence of a detergent, as a precipitate under ethanol, or all three!

10.2.2 Extraction of RNA

In the previous section we discussed the general principles behind RNA extraction, but how do we prepare RNA in practice? (You should be able to use your experience of DNA extraction to, at least partially answer this).

Here is our response:

- choose appropriate starting material (eg reticulocytes for the isolation of globin mRNA);

- use a procedure that gives good recovery and removes the contaminants described earlier.

There are three commonly used methods which all aim to dissolve nucleic acids effectively and remove protein and other contaminants as far as possible. The methods are:

- detergent and phenol extraction;

- guanidinium salt technique;

- protease K method.

We will discuss these in turn.

Detergent and phenol extraction

In the detergent and phenol extraction method the tissue is homogenised in a buffered concentrated detergent solution (commonly used detergents are sodium dodecyl sulphate, SDS, or tri-isopropyl-naphthalene sulphonate). This will dissolve most, if not all, cell components (plant cell walls, for example, would remain in the debris) but also has another important function. Can you guess what this might be?

As discussed earlier, the extraction process must inhibit ribonuclease. Detergents will do this by disrupting enzyme structure.

phenol
extraction

Proteins are removed by mixing the homogenate with an organic solvent, usually a solution of phenol or chloroform or a mixture of the two.

∏ What do you think will happen when proteins in the homogenate come into contact with phenol?

The proteins are denatured and will collect at the interface when the phenol:aqueous phases are separated. (See Section 7.2.3 if you do not remember this).

∏ Deproteinisation is carried out at the very start of RNA isolations. Why do you think this is necessary? Is it needed at the start of DNA isolations?

RNA would be degraded rapidly by RNases released from cells during lysis so the RNases must be inhibited as quickly as possible. There is no cheap, effective inhibitor of RNases that can be added to the extraction medium and so immediate deproteinisation is the usual way to destroy RNase activity. DNases can, generally, be inhibited completely by EDTA in the extraction medium and so deproteinisation can be carried out later and need not employ such powerful reagents.

∏ What does the aqueous phase contain?

Your answer should include RNA, DNA and low molecular mass molecules such as nucleotides, amino acids and sugars. There may be substantial amounts of carbohydrates but this is dependent upon the source from which the extract was prepared.

So how are these materials removed and the nucleic acids recovered?

The simplest thing to do is to exploit the high molecular mass nature of RNA and DNA. They are polar and will precipitate in appropriate organic compounds. Ethanol is added (approximately 2.5 times the volume of the aqueous phase) and the RNA and DNA will precipitate overnight at $4°C$, after a few hours in a freezer, or in about 20 minutes at $-70°C$. The nucleic acid pellet is recovered by centrifugation. Then it is washed in 70% aqueous ethanol at room temperature a few times, by suspension and centrifugation, to remove any traces of low molecular mass compounds and finally the nucleic acids are dissolved in water or a neutral buffer solution for further study.

Alternatively, nucleotides can be removed by gel filtration (Section 1.3.4); it is usually not necessary although it is often used to remove radioactively labelled nucleotides from nucleic acid preparations, for example in the production of labelled probes.

∏ Will the preparation be a pure RNA solution?

No, there will be DNA present which can be removed and possibly carbohydrate that can be difficult to remove. A frequently used technique is ultracentrifugation in a medium which exploits the difference in density between nucleic acids and contaminants. CsCl is often used for DNA (Section 1.2.3) or a concentrated solution of sodium acetate for RNA.

Guanidinium extraction

guanidinium method

Guanidinium thiocyanate is a strong denaturing agent that is effective in denaturing ribonuclease present in the sample. In this method the tissue is homogenised in a concentrated solution of guanidinium thiocyanate (5 mol l^{-1}) containing 3% 2-mercaptoethanol (a sulphydryl reagent, see Section 2.5). Debris is removed by centrifugation and the nucleic acids purified by centrifuging the solution at high speed through a dense solution, or 'cushion' as it is known, of caesium chloride, CsCl. The RNA forms a pellet, whilst DNA conveniently remains in the interface between the CsCl and the homogenate.

Protease K method

protease method

The protease method is a simple one: tissue homogenates containing detergent, usually SDS, are incubated at 65°C with protease K for half an hour or sometimes longer. The mixture is then deproteinised with phenol/chloroform as previously described.

∏ If the two methods just described remove protein by a chemical method why do some people wish to use another technique that involves an enzyme?

Some samples may contain protein which has a particularly high affinity for RNA. For example, many simple viruses, such as tobacco mosaic virus, have protein coats bound to RNA (most plant viruses contain RNA genomes).

∏ Are you surprised that the protease is incubated with a detergent?

It is a little unusual, but protease K is still active in the presence of SDS. The detergent will inhibit RNase (essential if a warm incubation of homogenate is involved) but denature most proteins whilst keeping them in solution. This makes them a better substrate for the protease.

10.2.3 Selection of the appropriate technique

The phenol extraction method is a generally applicable method. It is relatively quick and simple and works well on most tissues. The guanidinium salt method is very effective for extracting RNA from tissues rich in RNase; it is therefore widely used, particularly where the purpose of the experiment is to extract mRNA which is particularly susceptible to RNase attack (see Section 11.4) and must be obtained intact for the majority of experimental purposes.

The protease K method is unsuited to large-scale extractions where protein contamination is a problem, or as a second extraction following another method to remove any last traces of protein. The yield of RNA is approximately the same for each method but varies widely from tissue to tissue depending on cell size, growth rate and state of gene expression.

10.2.4 The removal of DNA

Π From what you have read so far, what approach can you suggest to remove DNA from a nucleic acid preparation?

density
gradient
centrifugation

If you remember the outline of the guanidinium thiocyanate procedure (Section 10.2.2) you will recall the CsCl centrifugation step. DNA has a lower buoyant density than RNA and will not sediment through a dense CsCl solution as easily as RNA. Thus CsCl density gradient centrifugation is one method of removing DNA. Alternatively, RNA and DNA have differential solubility in concentrated salt solutions (eg sodium acetate). However, this varies with the molecular mass of the RNA and therefore is not always a reliable method, particularly if a range of RNA molecules is required or if the size of RNA is not known.

The third and perhaps most widely used technique is to use an enzyme. It is now possible to obtain very pure DNase that is free of RNase contamination (or can be incubated with a protein RNase inhibitor, such as that obtained from human placenta, and is commercially available). Nucleic acid solutions are simply incubated with DNase in a neutral buffer solution containing Mg^{2+}. Removal of DNA can be monitored by gel electrophoresis (Section 1.3.4).

SAQ 10.1

What approaches are used to remove the following contaminants from a RNA solution:

1) protein; 2) lipid; 3) DNA; 4) nucleotides and nucleosides?

SAQ 10.2

What strategies are employed to avoid degradation of RNA by contaminating RNase activities?

10.3 The estimation of RNA

Estimation using UV spectroscopy

As we have already discussed in Chapter 7, nucleic acids have a marked peak absorbance in the ultra-violet region at 260 nm.

We can use adsorbance at 260 nm therefore as a measure of the amount of nucleic acids present. However, when we were discussing the estimation of DNA, we pointed out some of the problems and limitation of this approach. Can you remember what they were? If not, re-read Section 7.3, it will refresh your memory.

Π Is it possible to distinguish RNA and DNA from the scan in the UV region?

Unfortunately not, if the solution contains DNA an estimate of the concentration of RNA cannot be obtained. Therefore, we need to do some further work. Once approach is to remove the DNA first by one of the methods described in Section 10.2.4. Alternatively, there may be a chemical method to exploit the differences between RNA and DNA.

∏   What differences are there between RNA and DNA that could be used to quantify a specific nucleic acid?

The differences are in the ribose sugar and the presence of uracil or thymine.

∏   What would your choice be: to devise a test for the appropriate sugar or nitrogenous base?

The better approach is to use the sugar because if the base was used the measurement would depend not just on the amount of RNA but also on the nucleotide sequence.

Colorimetric estimation of RNA

There is a colorimetric test for RNA which uses the compound orcinol. The RNA is hydrolysed to nucleosides followed by conversion of ribose to furfural. Following the reaction with orcinol, the absorbance of unknown samples is read at 670 nm against known RNA standards used in a parallel reaction.

Hybridisation assay for RNA

∏   Are absorbance or colour reactions going to enable us to quantify specific RNA molecules such as a particular mRNA?

No: it is difficult to purify mRNA and even if it was achieved, the amount is likely to be very low. We use a different approach that involves more sophisticated experiments using labelled probes. These are either short oligonucleotides or longer molecules of DNA or RNA that have a sequence complementary to the RNA we wish to quantify. The probes are labelled with ^{32}P and used in hybridisation reactions followed by autoradiography. (Increasingly, non-radioactive labels are being used, such as enzyme co-factors, that can be detected as colour or light reactions).

We remind you of the process of hybridisation. Hybridisation works in the following way. When single-stranded nucleic acids are mixed together, under appropriate conditions, then hydrogen bonds can form between complementary strands. If long sequences of nucleotides on two strands of nucleic acids are complementary then many hydrogen bonds can be formed between the two strands. It will 'zip up', rather like a zipper clothing fastener to form a double standard nucleic and molecule (Figure 10.1). Such a molecule which contains two strands of nucleic acid from different sources is effectively a hybrid molecule - hence the term hybridisation. To detect hybrid molecules, often one of the strands is radioactively labelled with ^{32}P.

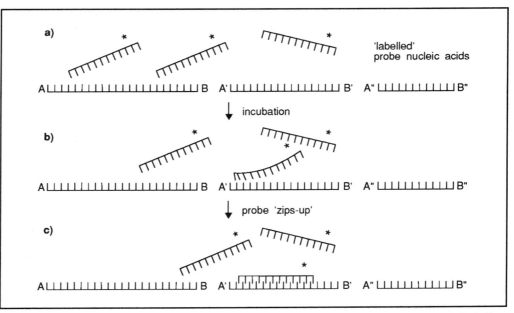

Figure 10.1 Stylised hybridisation assay using a labelled probe. Key: a) the mixture of mRNAs (A-B; A'-B'; A"-B") are incubated with labelled probe molecules; b) during incubation under suitable conditions, hydrogen bonds form between the probe molecules and complementary sequences on the mRNA; c) providing the nucleotide sequences are complementary, the probe forms a stable hybrid molecule with the mRNA. The probe and mRNA are held together by hydrogen bonds. By measuring the amount of label (*) held onto the mRNA, we have a measure of the amount of the mRNA present. In the case illustrated the probe used only has a complementary sequence to mRNA A'-B' and not to A-B and A"-B". The probe in this case is specific for a single type of mRNA and can, therefore, be used to measure this mRNA.

The simplest example is dot-blotting, where an unknown RNA sample is dotted onto a nylon membrane. This is then incubated with a solution of the labelled probe at a temperature suitable for hydrogen bonding of complementary sequences to occur. Following thorough washing of the membrane, autoradiography is used to detect the extent of hybridisation of the probe. The intensity of exposure of the X-ray film relates to the amount of target sequence.

SAQ 10.3

We have produced an ultra-violet absorption spectrum of a nucleic acid solution similar to that depicted in Figure 7.3. If adsorbance at 260 nm is 0.6:

1) What is the approximate nucleic acid concentration in μg cm^{-3}?

2) Which of the following statements best describes the solution tested?

   a) the solution contains DNA;

   b) the solution contains RNA;

   c) the solution contains DNA and RNA;

   d) the solution contains DNA and/or RNA.

3) What further test could you use to confirm your conclusions?

Summary and objectives

Three general methods are available for the extraction of RNA. Other cellular components are removed by a variety of simple procedures such as solvent extraction, enzyme degradation, selective precipitation and centrifugation.

Ribonuclease contamination is a potential problem when working with RNA but risks can be minimised with good experimental technique. Nucleic acid solutions can be analysed qualitatively (eg to monitor for protein contamination) and quantitatively by UV absorbance but it is difficult to distinguish DNA from RNA by this technique. Quantitative estimation can be carried out using a colorimetric assay in which hydrolysed RNA is reacted with orcinol.

On completion of this chapter you should be able to:

* describe the structure of RNA;

* outline the function of RNA molecules;

* understand the strategies for RNA extraction;

* outline three experimental approaches to RNA extraction;

* understand some of the techniques involved in obtaining a pure sample;

* calculate the concentration of nucleic acids in solution from absorbance data.

Purification of RNA

Purification of RNA

11.1 Introduction

In the previous chapter, concerned with RNA, we discussed extraction methods and the techniques employed in separating RNA from other components in the cell. In this chapter we will consider the approach and methods used to analyse RNA mixtures and to purify specific RNA molecules such as rRNA or specific mRNAs. After completing this section you will understand the purpose behind purification and have a general appreciation of the strategies and techniques used.

RNA molecules have varied functions in the cell, principally concerned with protein synthesis. It is only by studying individual types of RNA in detail that we can better understand their role and biological activity. Also many gene cloning strategies rely on the purification, or partial purification, of specific RNA as a first step.

Incidentally, specific RNA molecules are often referred to by the word 'species'. Obviously, RNA molecules cannot strictly be a species in the common use of the word within the biosciences; but the term is a useful one: for example, we may use the phrase 'a mRNA species'; meaning one particular mRNA with its unique coding capacity.

What does RNA purification involve? You can perhaps envisage two stages; RNA extraction, involving separation from other cell components, and then purification of the desired RNA species. Before we go any further, you could usefully revise some material covered in Chapter 10 by doing the following exercise.

SAQ 11.1	1) How do the methods for RNA extraction achieve separation of RNA from: a) proteins; b) lipids; c) small molecular mass compounds?
	2) What further procedures may be employed to remove carbohydrate and DNA?

11.2 The basis of purification

∏ In the spaces below, list three features of an RNA molecule that will distinguish it from other RNA molecules.

1)

2)

3)

You should have come up with the answers: 1) size; 2) base sequence; 3) secondary structure or shape. You should have been able to have listed 1) and 2) but you may not have thought of secondary structure.

∏ Which of the three distinguishing features listed above could form the basis of a purification method?

Size is probably the easiest to envisage, sequence seems a likely but a more complicated candidate and secondary structure possibly the most difficult. The various approaches and experimental methods used will be discussed in the next section.

11.3 Purification methods

In this section the various techniques that rely on size, or a combination of size and shape are considered.

11.3.1 Solubility

differential
solubility

In certain circumstances, high molecular mass RNA is less soluble than low molecular mass RNA. For example, high molecular mass RNA precipitates in concentrated solutions of sodium chloride (ie 4 mol l^{-1}) and can, therefore, be separated by centrifugation. Another example is that higher molecular mass nucleic acids will precipitate more readily in organic solvents. In ethanol precipitation; small molecular mass RNA will take several hours (overnight) to precipitate at -20°C, whereas large molecules will come out of solution within an hour or so.

∏ But what do you think is the inherent disadvantage of a solubility approach?

It is not very precise: it all depends on what we mean by high or low molecular mass.

What about the molecules in between, where is the boundary? For example, the technique is very useful for separating RNA polymers from constituent nucleotides. Ethanol precipitation in the presence of ammonium acetate separates RNA (precipitate) from nucleotides which stay in solution. Also differential solubility can be used as a crude method for separating high molecular mass RNA, such as the large ribosomal RNA from say transfer RNA. On the whole however, the technique is not a very useful one, particularly if the RNA species we wish to separate are all of high molecular mass (eg to separate mRNA from rRNA).

11.3.2 Density gradient centrifugation

During centrifugation in a suitable medium (Section 1.2.1), large molecules will sediment more quickly than small molecules. RNA can be purified in this way. The medium most commonly used is a sucrose solution.

∏ Make a list of some features that will be required of the centrifugation medium (eg the sucrose solution)?

It must, of course, be inert with respect to RNA; that is, it will not react with it or damage it in any way (Section 1.2.2). It must be pure so that it cannot contaminate the RNA; in particular, the sucrose solution must be free of ribonuclease. It must be of appropriate density to be capable of separating RNA molecules of different sizes, and be readily removed from the RNA when fractionation is complete.

density
gradient
centrifugation

In practice, the technique is a very simple one to execute. A gradient of sucrose concentrations is prepared in a centrifuge tube. This is most easily achieved by carefully pipetting successively more dilute solutions of sucrose on top of each other in a centrifuge tube, then leaving the tube in a cold room for a few hours or overnight for the layers to diffuse into a smooth gradient. Alternatively, the gradient can be made with a purpose-built gradient maker, consisting of two chambers containing a concentrated and a dilute sucrose solution connected by a tube (Figure 1.2). The exit tube is from the more concentrated solution and as the gradient is poured, the solution becomes gradually more dilute. (For more details see Section 1.2.2).

The choice of sucrose concentration and centrifugation speed will depend on the size range of RNA to be fractionated; for example, rRNA from eukaryotes can be fractionated using a 5-30% w/v solution centrifuged overnight at 80 000 g.

The RNA molecules will travel down the tube as zones and at rates dependent on the size and shape of the RNA. The composite effect of these two parameters on the rate of sedimentation is referred to as the sedimentation coefficient (s) measured in seconds or Svedberg units ($Sv \equiv 10^{-13}s$) For further details on centrifugation techniques and the mathematics involved see Section 1.2.3, and the BIOTOL text, 'Techniques used in Bioproduct Analysis'.

sedimentation coefficient

∏   How are we going to observe the RNA zones separating?

Following centrifugation, the gradient can be fractionated as depicted in Figure 11.1 and the fractions monitored by absorption of ultra-violet radiation.

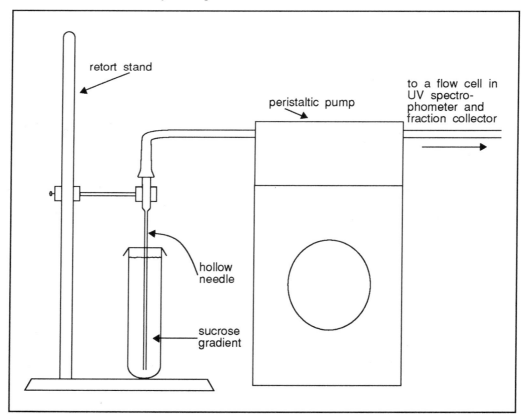

Figure 11.1 A method for fractionating a sucrose gradient. A more complex method can be arranged whereby a dense sucrose solution is pumped into the bottom of the gradient. An exit tube at the top is connected directly to a flow cell and, in this case, the gradient is therefore monitored from top to bottom.

| SAQ 11.2 | Can you remember the absorption maximum for RNA and the general shape of the ultra-violet spectrum? |

A typical absorbance profile from a density gradient centrifugation run carried out as described above is shown in Figure 7.2.

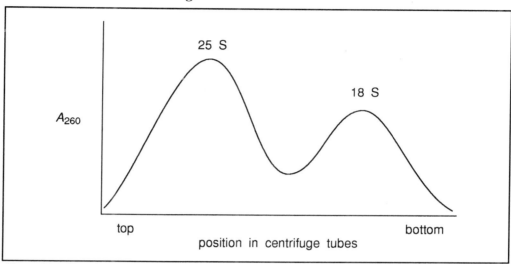

Figure 11.2 Absorbance profile of RNA extracted from dark-grown plants.

Π Look at the absorption profile in Figure 11.2. What are the peaks denoted 25 S and 18 S due to? The area under the 25 S peak is approximately twice that under the 18 S peak. Why is that?

The 25 S and 18 S RNAs are ribosomal molecules; each ribosome has one of each, the 25 S being a component of the large subunit and the 18 S being present in the small subunit. The 25 S RNA is nearly twice as long as the 18 S (relative molecule masses, RMM 1.25×10^6 and 0.7×10^6, respectively) and will, therefore, absorb nearly twice the amount of ultra-violet radiation for the same number of molecules.

Π If the 25 S molecule is about twice the size of the 18 S molecule why do the values for the sedimentation coefficients not similarly correspond?

Because the sedimentation coefficient depends on a combination of size and shape; a molecule that is twice as long but also folded will not have a sedimentation coefficient that is twice as large. The same effect is seen in the analysis of ribosomes. An intact ribosome obtained from the cytosol will have a sedimentation coefficient of 80 S. It is made up of a large subunit having a sedimentation coefficient of 60 S, and a small (40 S) subunit. Here you see that sedimentation coefficients are not additive.

Π What is your opinion of the resolution obtained in Figure 11.2?

Yes, it is rather poor isn't it? Sucrose gradients are a useful preparative technique because quite large amounts of RNA can be fractionated, but the method is not a good analytical one because of the poor resolution. Agarose gel electrophoresis gives much improved resolution and has superseded sucrose gradients as an analytical technique.

11.3.3 Gel electrophoresis

RNA is a polyanion and will, therefore, migrate to the positive electrode during electrophoresis. If this occurs through a gel of appropriate pore size, the mobility (ie rate of movement per unit applied field) of the RNA molecules relates to the logarithm of their RMM; the smallest molecules migrate at the greatest rates because of the sieving effect of the gel. (For more details see Section 1.3.4).

gel
electrophoresis

The technique is most commonly carried out using agarose in a horizontal, submerged gel electrophoresis system. The method is essentially the same as for DNA. Therefore, refer to the chapter on DNA fractionation (Chapter 8) and read Section 8.3.1 on DNA gel electrophoresis, before continuing with this section, (only the modifications required for RNA fractionation are discussed here). So, assuming you have read this material, try SAQ 11.3.

| SAQ 11.3 |

1) A restriction enzyme digest of bacteriophage λ is a frequently used as RMM reference for DNA gel electrophoresis. What is the structure of the λ fragments (are they circular, linear, supercoiled)?

2) If supercoiled, open circle and linear DNA molecules of identical RMM are run on an agarose gel, what are the relative mobilities?

RNA can be fractionated in an agarose gel in a way similar to that used for DNA, ie same apparatus, buffer solution, etc. The only difference is that a slightly higher agarose concentration is normally used (1.1% instead of 0.8%) more appropriate to the relevant RMM.

∏ In this kind of system do you expect RNA mobility to be solely related to RMM?

The answer is no. Just as with agarose gel electrophoresis of DNA, the system maintains H-bonds. RNA molecules will, therefore, maintain their secondary structure. The consequence is that RNA molecules of the same RMM may not co-migrate, just as supercoiled and linear DNA molecules with the same number of base pairs have different mobility values. To ensure that mobility is related to size, a denaturing system is required. Such gels can be used to determine the RMM of RNA (Section 8.3.2).

11.3.4 Chromatography

Various chromatographic techniques can be employed to fractionate RNA. The choice of method will depend on the purpose behind the fractionation and can exploit the size difference in RNA molecules, the negative charge of RNA, or differences in nucleotide sequence.

Size

∏ What type of chromatography exploits the differences in size of molecules?

gel exclusion
chrom-
atography

The technique is called gel exclusion chromatography (Section 1.3.4).

It is not widely used for RNA fractionation but having said that, it is routinely used for separation of nucleic acids (DNA or RNA) from nucleotides. In particular, radioactive nucleotides, used as substrates for *in vitro* labelling techniques, often need to be separated from DNA and RNA molecules. Labelled nucleic acids can be used as 'probes'; that is, they can be employed in nucleic acid hybridisation experiments where they will base pair with complementary target sequences. We will discuss the application of this technique for identifying RNA (or DNA) molecules in a little more detail in the next chapter.

The separation is based on quite a simple principle: a chromatographic gel contains pores; molecules that are large relative to these pores will not enter the gel whereas smaller molecules will. Therefore, the smaller molecules will be retarded and consequently be separated from larger ones. In practice, RNA and DNA can be separated from nucleotides using a small gel filtration column run under gravity. Alternatively, the gel and buffer solution can be poured into a plastic tip of the type normally used on automatic pipettes. If a mixture of nucleic acids and nucleotides is then added, and the tip is put inside a centrifuge tube and centrifuged, the nucleic acids will elute along with the buffer solution, but the nucleotides will be left behind in the gel remaining in the tip. Not surprisingly, these are called 'spin columns'.

Charge

RNA is highly charged and will bind to anion-exchange resins such as DEAE-cellulose. Binding is very strong and is directly proportional to RMM. The technique is therefore most appropriate for the fractionation of low molecular mass RNA and has been used for the analysis and purification of tRNA. The principle of ion-exchange can be used in high-performance liquid chromatography (HPLC) of nucleic acids (Section 1.3.3). HPLC has traditionally been used for the separation of small molecules eg nucleotides, rather than macromolecules, but the scope of HPLC has increased in recent years because of the development of the technique, for example in the design of stationary phases with higher flow rates and particle sizes. HPLC has, however, not been very widely used for RNA separation because of the high resolution available with gel electrophoresis. An ion-exchange separation of RNA is shown in Figure 11.3. illustrating its use to assay for the presence of viral RNA in a plant extract.

ion exchange chromatography

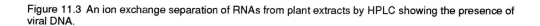

Figure 11.3 An ion exchange separation of RNAs from plant extracts by HPLC showing the presence of viral DNA.

Sequence

Affinity chromatography is a technique frequently used in the purification of proteins (Section 1.3.4), and has been applied to RNA.

∏ What is meant by the term 'affinity chromatography'?

Separation is based on a particular feature of the molecule(s) you wish to purify. A chromatographic system is devised such that the required molecule binds to an insoluble support. Contaminating material is washed away and the required molecules eluted off. The technique has frequently been used for the purification of mRNA obtained from eukaryotes.

∏ What feature of eukaryotic mRNA makes it particularly suitable for affinity chromatography?

You may not know the answer to this but the vast majority of mRNA molecules in eukaryotes have a poly(A) tail at their 3' ends. This feature can be used to design an affinity chromatography purification technique for mRNA. This is especially valuable as we are often interested in isolating mRNAs because of the 'information' they contain.

∏ What kind of molecule would you attach to an insoluble support to create an affinity chromatography column for poly(A)-containing RNA?

The simple answer is to take advantage of the base-pairing rules. If we attach some poly(T) or poly(U) to the column, the poly(A) tails of the mRNA will hydrogen bond under appropriate conditions. It is as simple as that. The poly (A)$^+$ RNA will hybridise (ie bind) to the column material (usually oligo(dT)-cellulose or poly(U)-Sepharose) at neutral pH and in the presence of a salt such as sodium chloride or sodium acetate. This salt is necessary to neutralise the negative charges on the phosphate groups and therefore stabilise the hybrid. If the column is washed thoroughly in the same buffer solution, nucleic acids not containing tracts of poly(A) will be removed.

∏ How can we recover the poly(A)$^+$RNA?

Destabilise the H-bonds. This is most easily achieved by washing the column with a low salt buffer solution. The mRNA can therefore easily be eluted from the column.

11.4 The isolation of specific mRNA.

In the early years of gene cloning, obtaining specific pure mRNA was an important step in isolating and purifying a gene. Let us take the globin genes as an example, as they were some of the first human genes to be isolated. Poly(A) RNA was obtained from red blood cells where globin mRNA was present as a large proportion of the total mRNA. The globin mRNA was purified by sucrose gradient centrifugation and then used as a template for the synthesis of copy DNA (cDNA). Copy DNA is DNA that has been synthesised using RNA-dependent DNA polymerase (also known as reverse transcriptase), nucleotide triphosphates and an RNA template (see Chapter 6). The cDNA made by this procedure was used to probe a human genome library and led, subsequently, to the isolation of the globin genes. The cDNA step is useful because DNA molecules can be cloned within plasmids in bacteria, so large amounts of a particular sequence, in this case the globin coding sequence, can be generated. The fact that it is cDNA not mRNA does not matter, it contains the same sequence (in double-stranded form) and is, therefore, a useful probe to locate the chromosomal gene.

∏ If the cDNA has the sequence to code for globin why do we not refer to it as the globin gene?

coding sequences

Coding sequences are only part of the true fully-functional gene. Therefore, we use the coding sequence, as in cDNA, to probe a genome library to obtain the complete gene, including introns and surrounding regulatory elements. If you are unfamiliar with the terms 'introns' and 'regulatory elements', you are recommended to read the BIOTOL text, 'The Infrastructure and Activities of Cells'.

Let us come back to the RNA again. In general terms, how easy do you think it is to purify a mRNA species on the basis of its size (as was done for the globin mRNA)? On the face of, it you might suppose it is quite easy, but it is not.

∏ Try to think of two reasons for this.

• Different mRNA coding for different polypeptides may have the same approximate size;

• some mRNA may be present in small amounts and therefore be difficult to detect.

Which leads on to explaining something not yet mentioned. How do you think we can detect and identify mRNA? Measuring the A_{260} will only tell us about the concentration of nucleic acid, not the type; we need something more specific.

In vitro protein synthesis

in vitro protein synthesis

The technique missing is in vitro protein synthesis. This is a technique where a relatively crude cell extract is used to translate any exogenous mRNA that is added to it. Some form of energy (ATP, GTP) is supplied and all twenty amino acids are added, usually including a radioactive one, such as ^{35}S-methionine. The cell extract contains ribosomes, tRNA, the amino-acyl-tRNA synthetases (these add amino acids to specific tRNAs) and all the other factors required for polypeptide synthesis. The type of proteins made will depend on the added mRNA.

Of course, the technique relies on a system for identifying the protein made. The methods are: SDS PAGE, Western blotting and immunodetection, in which antibodies are used to 'identify' specific proteins.

Going back to the case of globin again, the mRNA was fractionated on sucrose gradients and identified by in vitro protein synthesis. That is, fractions from the gradient were tested, in turn, to see which one could code for globin proteins. However, the approach used for globin is not always feasible. It does not, for example, enable the purification of a mRNA that is not particularly abundant. In this case other techniques can be tried. One approach has been to treat polyribosome preparations with antibodies to the protein coded by the mRNA wanted. This causes precipitation of those ribosomes containing the relevant nascent proteins. The mRNA can then be extracted from this enriched preparation.

SAQ 11.4

1)   What is an *in vitro* protein synthesis system and what is it used for?

2)   What are the components of the system?

But, at this point, let us step back for a moment, and go over some arguments again.

∏   Why do we wish to purify specific mRNA?

The reasons are most likely to be:

*   to use it to make a probe to isolate a gene;
*   to study gene expression during, say, tissue development, or both.

In other words, mRNA isolation is usually a means to an end and not the end in itself. So can the mRNA purification step be bypassed altogether?

Yes. It is still necessary to isolate mRNA, and separate it from rRNA and tRNA by affinity chromatography, but there is currently far less need to purify individual mRNA. This is because of the variety and power of genetic manipulation strategies and, more recently, the polymerase chain reaction.

11.5 The use of cDNA

We have already outlined the process of genetic engineering in Chapter 6. It is such an important process that it is worthwhile developing this discussion further here especially related to cDNA.

To keep things simple, let us stick to the globin example and imagine how we currently might go about producing a probe to isolate the chromosomal gene.

We will use the red blood cell again; but this time simply isolate total mRNA and we do not worry about trying to purify the globin mRNA. Having obtained this total mRNA, DNA copies of the entire mixture can be made. This is called a cDNA library. It is a collection of DNA copies of all mRNAs from red blood cells. In fact it is a red blood cell cDNA library.

genetic engineering

This mixture of cDNAs can be inserted into a plasmid and replicated in bacterial cells using the techniques of genetic engineering. For this small, circular pieces of DNA called plasmids, are 'cut' open using a restriction enzyme and then incubated with the cDNAs in the presence of an enzyme, DNA-ligase, capable of joining the ends of the DNA together. In this way the plasmids may reform carrying cDNA as shown in Figure 11.4.

This can be arranged in such a way that individual colonies of bacteria growing on a plate only contain one cDNA, but the colonies collectively, ie on the whole plate, contain the complete cDNA library (see also Chapter 6).

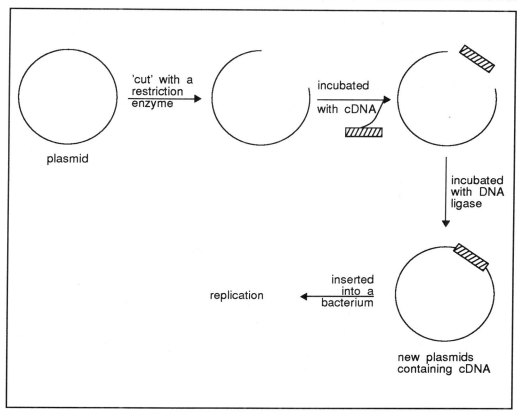

Figure 11.4 The principal steps in making copies of cDNA.

The question that needs to be addressed now is which colony or colonies on the plate contain the globin cDNA?

To find out, we need to characterise the DNA in the colonies. We can take a colony and extract its DNA. The DNA is then denatured by boiling or with alkali treatment, and the single-stranded DNA bound to a membrane filter. Now, imagine that some total mRNA from red blood cells is poured through the membrane filter under conditions that favour nucleic acid hybridisation.

∏ What do you think will happen?

If the DNA bound to the filter contains cDNA for globin, the globin mRNA will hybridise.

∏ But how do we know if this has happened?

This bound mRNA is eluted off and translated in an *in vitro* translation system. If globin protein is made we have the right cDNA. For *in vitro* protein translation we will, of course, have to put in all of the ingredients needed for protein synthesis described earlier.

We hope you have realised that you need to keep a replica of the original plate! Because we have used the colony to extract the cDNA we need the replica plate. From this we can obtain the relevant colony and therefore more of the cDNA.

If the colony was found not to contain globin cDNA, the process is repeated with other colonies until the cDNA is found. But, those of you with a practical bent may be asking the following question: there may be many colonies growing on the original plate; hundreds or possibly thousands; surely, it is very laborious to test them all by this approach?

Yes, but you can arrange it so that you take a number of colonies at one go, say from an eighth of the plate, and test them in batches. When you get a positive result for a batch, work your way down to the right colony - again because you have kept replica plates.

But let us see if you can remember a fact we introduced earlier.

∏ If you were to pick a colony at random from a cDNA library of red blood cells what do you think the chances will be that the cDNA would contain the globin sequence?

Probably about 50:50. Do you remember? We said about half of all the mRNA in red blood cells is globin mRNA. Therefore, approximately half of the colonies on our plate should contain globin cDNA. So, you see, it probably is not so difficult after all.

In fact, it gets easier. Keeping to the same globin example, consider a replica of the red blood cell cDNA library on a membrane filter prepared as shown in Figure 11.5.

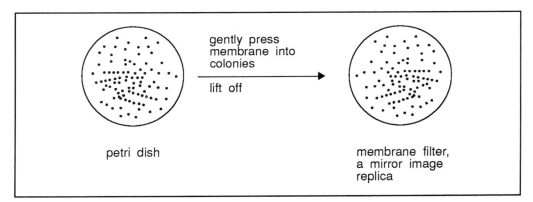

Figure 11.5 Replication of cDNA library.

All the cells on the membrane can then be lysed and the DNA denatured (this can be done in such a way that DNA denatures and sticks without smudging).

The membrane is now floated in a solution of radioactive total cDNA prepared from red blood cells, in other words, the same cDNA that was put into plasmids to transform the colonies in the first place, and nucleic acid hybridisation allowed to take place. Finally the membrane is washed to remove any unbound labelled cDNA and the filter exposed to autoradiography.

SAQ 11.5	The circles in Figure 11.6 represent the filter (with less colonies than shown above to make your task a little easier) and two autoradiograms. Black dots will appear on the X-ray film following development. As the X-ray film is exposed to the filter the signals will get darker and more dots may be apparent.

1) Try and fill in the dots where you think they might appear. To help you, think about the probe. It is a mixture of cDNA made from mRNA from red blood cells.

2) What proportion of this mRNA codes for globin?

3) What proportion of the colonies on the filter contain cDNA copied from the globin mRNA?

4) Will the spots of cDNA become equally radioactive upon hybridisation to the labelled probe and therefore light up the autoradiogram at equal rates?

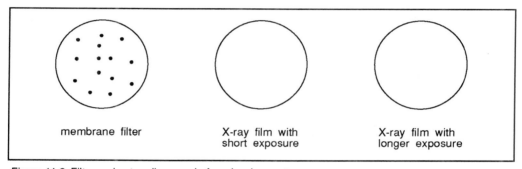

Figure 11.6 Filter and autoradiograms before development.

SAQ 11.6	Complete the following table which relates to the different methods of fractionation of RNA.

	Fractionation method	Advantages	Disadvantages
1)	solubility		
2)	sucrose gradient frationation		
3)	agarose gel electrophoresis		

4)	chromatography
	  gel exclusion
	  ion exchange
	  HPLC
	  affinity		

Summary and objectives

There are a number of different approaches for the fractionation of RNA. They all have their relative advantages and disadvantages.

Gene cloning strategies have, to an extent, replaced the need for highly purified mRNA. This advanced strategy is described fuller in the BIOTOL text, 'Techniques for Engineering Genes'.

On completion of this chapter you should now be able to:

- provide a list of the types of RNA in cells;

- explain the basis of RNA purification;

- describe several purification methods;

- discuss their relative advantages and disadvantages;

- realise the importance of mRNA isolation;

- understand the strategies involved in purifying a mRNA coding sequence;

- realise the value of cDNA cloning.

Determination of the chemical composition, structure and physico-chemical properties of RNA

Determination of the chemical composition, structure and physico-chemical properties of RNA

12.1 Introduction

The aim of this final chapter is to explain how we characterise RNA molecules. This knowledge is of value if we are to understand more about the mechanism of RNA synthesis and its role within the cell. After working through this chapter you will know more about the techniques involved in studying polymer size, nucleotide sequence and shape and understand how this knowledge is applied to the study of RNA function.

This chapter has been written on the assumption that the reader has a basic knowledge of protein synthesis and the roles of RNA in this process. If the reader feels uncertain of this process, the BIOTOL text, 'The Infrastructure and Activities of Cells', contains more details. The present chapter provides a bridge between a basic understanding of protein synthesis and the more advanced treatment of nucleic acids needed to understand regulation of gene expression and genetic manipulation. These latter two aspects are dealt with in the BIOTOL texts, 'Genome Management in Prokaryotes'; 'Genome Management in Eukaryotes'; and 'Techniques for Engineering Genes'.

The principal role of RNA is to act as an intermediate in gene expression. To learn more about this role we need to understand the mechanism of RNA synthesis, and the precise structure and function relationship for each RNA species. The first two steps, extraction and purification, were discussed in Chapters 10 and 11. In this chapter we will go through various techniques used in RNA analysis.

Perhaps the simplest characteristic of an RNA molecule to determine is its relative molecular mass (RMM): we will discuss that first. Subsequently, it is necessary to learn about its base composition and nucleotide sequence. That is the key to establishing the coding capacity, if we are interested in mRNA, and potential secondary and tertiary structures. Finally, some examples of how structure and chemical properties relate to function will be considered.

12.2 The determination of the relative molecular mass of RNA

Π   Can you remember which of the fractionation techniques (described in Section 11.3.3) was based on size of RNA and provided good resolution?

gel electrophoresis   It was agarose gel electrophoresis.

But was fractionation entirely dependent on size?

No, secondary structure also played its part and it was concluded that the separation of RNA molecules would have to be carried out under denaturing conditions if the mobility value was to more accurately reflect the length of RNA.

∏ What do we have to do to denature nucleic acids and how is it most easily achieved?

To denature nucleic acids we have to break interactions that hold intact the 3-D structure, so we will have to break H-bonds.

The easiest ways to achieve this are by heat or with alkali treatment. But these methods are not appropriate to agarose gel electrophoresis. Why? Because, quite simply, they are impractical: the temperature required to denature nucleic acids and keep them single-stranded will melt the agarose and concentrated alkali is not a good environment for electrophoresis.

Instead a different, chemical approach is used. One method is to include formamide in the buffer solutions, as this compound destabilises hydrogen bonds and effectively lowers the melting temperature of nucleic acids.

Alternatively, RNA can be treated with glyoxal (ethanediol) and dimethyl sulphoxide (DMSO) at 50°C. The DMSO disrupts H-bonds and the glyoxal modifies guanine residues to prevent re-annealing (Figure 12.1). Electrophoresis is then carried out in a low ionic strength buffer solution to help maintain the RNA in a linear form. A feature of the glyoxal technique is that native (that is non-glyoxal treated) and denatured nucleic acids can be run together on the same gel.

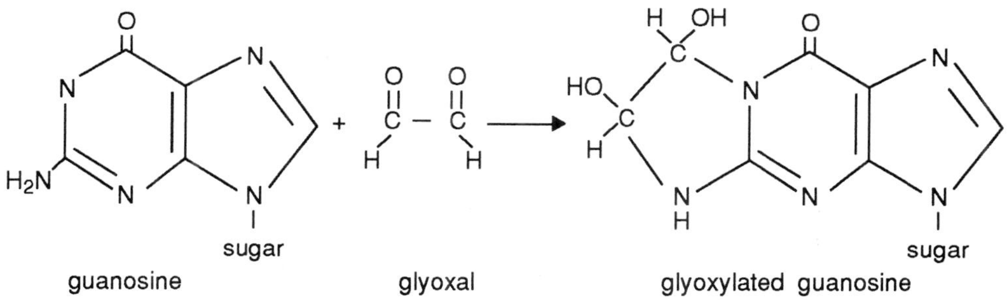

Figure 12.1 The reaction between glyoxal and guanosine.
(Carmichael, SS & McMaster, GK (1980), Methods in Enzymology, 65, 380-391).

∏ Would you expect native and denatured RNA of the same RMM to have the same mobility and if not which will run more slowly?

No, denatured RNA and native RNA will not have the same mobility. The denatured RNA will run more slowly for the same reason that linear DNA migrates more slowly than supercoiled DNA. It is less compact.

Following electrophoresis, the gel can be stained with a dye such as ethidium bromide (Figure 12.2). However, we cannot expect a fluorescent dye such as this to stain denatured nucleic acids very well as it acts by interchelating between nucleotide pairs.However, this is not a problem as the glyoxylation reaction is readily reversed by soaking the gel in 0.3 mol l^{-1} NaOH. The gel can then be stained and bands viewed with an ultra-violet transilluminator.

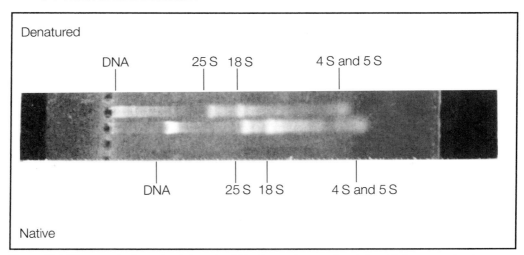

Figure 12.2 Photograph of a stained glyoxal gel. The lanes show total nucleic acids from pea shoots grown in the dark: on the bottom, glyoxylated; on the top, non-glyoxylated. The numbers 25 S, 18 S and 5 S refer to sizes of rRNA; 4 S is tRNA.

Because the RNA has been fractionated as single-stranded molecules, the migration will be related to the RMM. We can draw a graph of mobility against log RMM just as for DNA gels (Section 8.2). Ribosomal RNAs from different sources (eg 28 S and 18 S from animals; 25 S and 18 S from plants and 23 S and 16 S from bacteria or chloroplasts) can be used as RMM markers.

∏ Will staining of gels be the most appropriate technique for determining the RMM of mRNA separated by denaturing gel electrophoresis?

No. Individual mRNA molecules are unlikely to be present in sufficient quantity to be detected; the staining is simply not sensitive enough, and even if it was we would see a smear rather than discrete bands because of the number and variety of sizes of mRNA molecules.

Southern blotting

The problem can be solved using the technique of Northern blotting. Can you recall Southern blotting? It is the technique, devised by Ed Southern, for identifying DNA sequences separated on agarose gels. In this technique, a nitrocellulose or nylon filter is pressed onto the agarose gel after electrophoresis. Part of each of the DNA sequences will stick to the filter. When these filters are treated to fix the DNA, the result is a replica of the agarose gel electrophoretograms. The DNA attached to the filter can be screened by DNA-DNA hybridisation. The uptake of DNA from the gel by the membrane is a process of capillary transfer and is best carried out in 2 x SSC (SSC = standard saline citrate, ie a 0.15 mol l^{-1} of sodium chloride and 0.015 mol l^{-1} sodium citrate). (See Chapter 6).

Northern blotting

Northern blotting is exactly the same principle except that RNA is transferred from a gel, not DNA. The only differences from the method for DNA is that a higher concentration of buffer solution (20 x SSC rather than 2 x SSC) is used for the capillary transfer, to assist RNA binding to the membrane. Incidentally, the technique was not devised by anyone called Northern; it is a sort of joke!

So how do we locate and identify the mRNA stuck to the membrane (referred to as a Northern blot)? A hybridisation reaction with a probe is required. If an appropriate labelled cDNA is denatured and incubated with the Northern blot under conditions that permit nucleic acid hybridisation, the probe will hybridise to the target mRNA and be revealed by autoradiography following agarose gel electrophoresis.

The position of the band or bands on the autoradiogram can then be compared with the position of the known markers and the RMM determined, if desired, by reference to a graph of log RMM against distance moved.

Northern blotting has frequently been used to determine the size of mature mRNA molecules but it can also be used to study mRNA precursors. We know that mRNA molecules are synthesised as large precursors in the nucleus. If nuclear RNA is fractionated on agarose gels and investigated by blotting and hybridisation, bands are revealed that represent larger molecular mass species than mature mRNA isolated from cytoplasm.

Another use of Northern blotting is to study changes in gene expression as a result of, for example, hormone action or cell differentiation during development. RNA is extracted at different times following hormone treatment, or at different times during cellular development. The appearance or disappearance of bands on a Northern blot probed with a specific cDNA will show changes in mRNA levels. Of course it is not always necessary to determine the RMM on these occasions, it may already be known. In this case a simpler technique known as dot blotting can be used. This simply involves spotting samples of RNA onto a membrane and assaying for the amount of specific mRNA by measuring the extent of hybridisation of a cDNA by comparing the intensity of dots on the X-ray film against known standards.

| SAQ 12.1 | 1) Why is a denaturing gel electrophoresis system required to measure the RMM of RNA? |
| | 2) Name a denaturing agent and explain the effect of denaturation on the mobility. |

| SAQ 12.2 | A research group wishes to know the RMM of a mRNA that is present in only small amounts in the cells they are investigating. How do they achieve their objective? |

12.3 The analysis of the base content of RNA

12.3.1 Base composition

The nucleotide composition (also referred to as the base composition) is a characteristic of an RNA molecule, and it is usually expressed as %A, %G, %C and %U. There may also be some modified nucleotides.

∏ What steps do you think are involved in determining the base composition of an RNA preparation?

First of all we need to break down the polymer into its constituent nucleotides.

alkaline
hydrolysis This can be done by alkaline hydrolysis with sodium hydroxide, depicted in Figure 12.3, or by digestion with RNase T$_2$, an endonuclease that is not base specific.

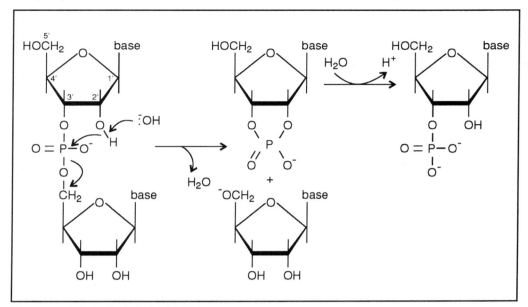

Figure 12.3 Alkaline hydrolysis of RNA.

Second, we need to analyse the mixture. This can be achieved by HPLC (Section 1.3.3) or two-dimensional TLC (Section 1.3.3), (using isobutyric acid and ammonia as solvent for the first dimension, and isopropanol and HCl for the second) against known standards. Identification of the nucleotides is most readily achieved by carrying out the analysis on ^{32}P-labelled nucleotides. Therefore, prior to analysis, the RNA digestion products are made radioactive by treatment with polynucleotide kinase and 5'-[γ-^{32}P]ATP.

nucleotide
analysis In the early days of molecular biology, nucleotide analysis was an important technique for characterising RNA molecules; but that was before the advent of gene cloning and more streamlined sequencing techniques. However, base analysis reveals the presence of modified nucleotides in RNA; these are an important feature of RNA and contribute to the correct functioning of the molecules. A full description of the variety and function of all modified nucleotides is beyond the scope of this chapter but a brief mention of some examples is given below.

When RNA is analysed, it is found that many of the nucleotides are methylated on the 2' hydroxyl of the ribose units especially in rRNA. That is, the group 2'-OH is replaced with 2'-OCH$_3$.

∏ The catalytic site of ribonuclease mimics alkaline hydrolysis as shown in Figure 12.3. What, therefore do you think is the effect of the methylation?

The reaction between the 2'-OH and the OH$^-$ cannot take place and the 2',3'-cyclic nucleotide cannot form. Therefore methylated RNA is more resistant to ribonuclease attack.

This makes sense. Ribosomal RNA needs to be stable in the cell to form ribosomes for protein synthesis. Messenger RNA, on the other hand, is not necessarily required to be stable because the cell needs to be able to cease synthesis of specific proteins to respond to environmental or developmental pressures.

A second example of even more extensive modifications is found in transfer RNA. More than fifty modified nucleotides have been identified. They are not all found in all species or in all tRNA, but some, such as ribosylthymidine (T), pseudouridine (ψ) and dihydrouridine (D) are found in tRNA molecules of most organisms.

∏   What do you suppose is the function of all these modified nucleotides?

It would seem that the presence of modified nucleotides is responsible for the specific shape of tRNA molecules by, for example, preventing certain regions from hydrogen bonding. They are also responsible for many of the specific functions of the tRNA molecule.

The most readily understood function of modified nucleotides can be seen in the anticodon. Modifications allow a greater range of base-pairing to occur and are partly responsible for the chemical basis of the 'Wobble hypothesis'. For example, inosine is often present as the first base in the anticodon. It is able to base pair with U, C or A (Figure 12.4). In other words there is often not a stringent requirement for a particular nucleotide in the first position in the anticodon. This lack of stringency allows some 'wobble' on the nature of the complementary (third) base in the code. A diagram depicting the secondary structure of tRNA, showing the location of the more common modifications is shown in Figure 12.5.

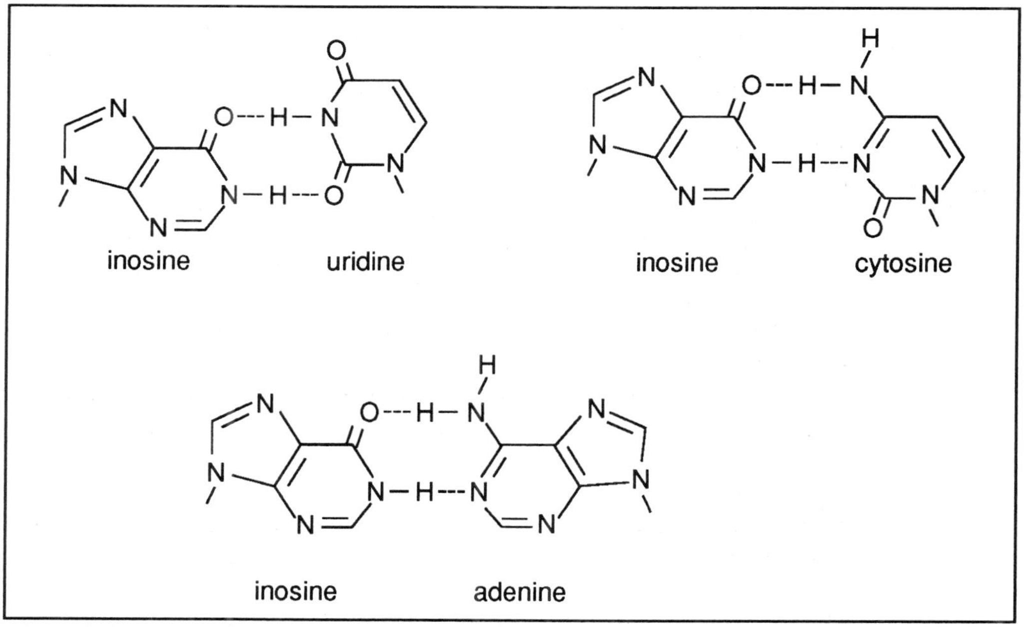

Figure 12.4 Base pairing of inosine.

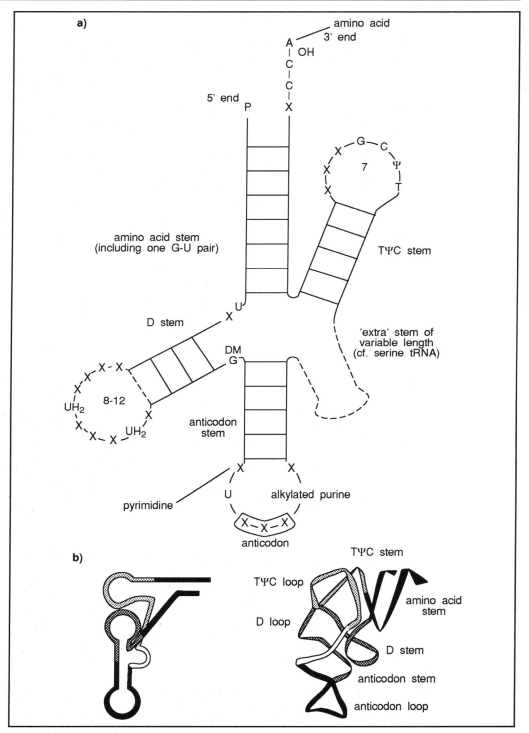

Figure 12.5 The structure of tRNA. a) Secondary structure showing base pairing and commonly found positions of some modified nucleotides (X = any nucleotide). b) Tertiary structure: left a projection showing how double helical regions form at right angles; and right, a schematic drawing of the three-dimensional structure.

12.3.2 Nucleotide sequencing

Although knowledge of the base composition is valuable, much more can be learned about RNA molecules if we can determine the full sequence. The first nucleic acid ever to be sequenced was the alanine tRNA of baker's yeast consisting of 77 nucleotides. This was achieved by Holley and coworkers in 1965. About 149 kg of baker's yeast was required to isolate 1 g of the tRNA; after nine years of effort the sequence was established and Holley received the Nobel prize in 1968. The strategy was similar to that used to determine the amino acid sequence of polypeptides (Section 4.3.1).

First, the RNA was cut into smaller fragments using base-specific endonucleases such as those shown in Table 12.1. The resulting oligonucleotides were separated in two dimensions by electrophoresis and chromatography.

Enzyme	Substrate	Specificity	Cleavage site
micrococcal nuclease	RNA and DNA	for adenine and non-specific base (X) on the 5' side	....Xp↓Ap....
takadiastase T1	RNA	for guanine and non-specific base on the 3'side	....Gp↓Xp....
pancreatic ribonuclease A	RNA	for a pyrimidine (Py) and a non-specific base on the 3' side	....Py↓Xp....

Table 12.1 Characteristics of some endonucleases.

These were eluted and individually investigated with further enzymes, for example by digestion for different times with exonucleases (for example: snake venom nuclease, which attacks from the 3'-OH end to yield nucleoside 5' monophosphates, and spleen nuclease, which cleaves from the 5' end, although it requires the 5' phosphate to be removed first, to yield nucleoside 3' monophosphates). The digestion products were identified as discussed in the previous section. The whole exercise was repeated, using different nucleases to generate a new pattern of oligonucleotides. Eventually the whole sequence was obtained by a series of logical arguments based on overlapping sequences. Subsequently, more ambitious projects were embarked on, including the report in 1976 of the complete sequence of the MS2 bacterial virus; a RNA of 3569 nucleotides.

RNA sequencing

The chromatographic approach to RNA sequencing has largely been superseded by techniques which employ high resolution polyacrylamide gels, but the first stage, that is the separation of oligonucleotides, is still used. The pattern of spots obtained is referred to as an RNA 'fingerprint' and is characteristic for an individual RNA. The technique can be used where knowledge of the complete sequence is not required. For example, to compare rRNAs of closely related species or to study evolution or epidemiology of RNA viruses. Another application is to identify base modifications: if this has occurred at the site of cleavage by a specific nuclease, the fingerprint will be changed.

Currently, RNA is sequenced using the same basic strategy as that used for DNA. In fact, there are so many things in common that it is not necessary to go through the entire RNA sequencing protocol in detail. Rather the approach will be to point out the differences between the experimental techniques. If you are unfamiliar with the overall approaches to DNA sequencing, that is the chemical cleavage and dideoxy nucleotide methods, then re-read Sections 9.6.1 and 9.6.2 before continuing with this section.

Alternatively, if you feel fairly confident about the concepts behind DNA sequencing, read on, for this will act as revision as well as teaching you about RNA sequencing.

Consider the chemical cleavage approach first. Can you remember the overall strategy for this approach, devised for DNA by Maxam and Gilbert?

The same approach can be used for RNA. Firstly the polymer is end-labelled, at either the 5′ or 3′ end. The former is achieved by dephosphorylating the RNA with alkaline phosphatase, followed by $5'^{32}P$ end-labelling with polynucleotide kinase and $5'$-$[\gamma$-$^{32}P]ATP$. The 3′ end can be labelled either with $[5'$-$^{32}P]pCp$ and RNA ligase or with $5'[\alpha$-$^{32}P]ATP$ and poly(A) polymerase. Following end-labelling, the sample is divided into four portions and chemically modified at specific bases as shown in Table 12.3. The RNA is then cleaved with aniline under sub-optimal conditions, such that a mixture of fragment lengths is produced by all of the four reactions. Of course, all fragments in a specific portion will be cleaved at the same base. These four product mixtures are loaded onto four separate lanes of a polyacrylamide gel, and subjected to electrophoresis and autoradiography to separate and observe the fragments. The sequence is read just as for DNA.

Nucleotide	Modification reaction
G	dimethyl sulphate in NaOAc at 90°C; then reduction with NaBH₄
A	diethyl pyrocarbonate in NaOAc at 90°C
U	50% hydrazine on ice
C	NaCl and hydrazine on ice
(NaOAc = sodium acetate)	

Table 12.2 Reactions for chemical modifications of RNA prior to cleavage with aniline.

The end-labelling and specific base cleavage approach can, alternatively, be carried out using base-specific partial enzymatic digestion in place of chemical cleavage. The enzymes used are shown in Table 12.3.

Specificity	Nuclease
Gp/X	RNase T1
Cp/X	RNase CL3
Xp/U and Xp/A	nuclease S7
Ap/X	RNase U2
Xp/X(ie random)	sulphuric acid
(X = any nucleotide)	

Table 12.3 Sequencing with base-specific partial enzymic digestion.

Note that nuclease S7 cleaves at the 5′side of both U and A. Consequently, because we have labelled at the 5′ end the band on a gel corresponding to a U or A, cut by S7, will be one nucleotide shorter than fragments specifically cleaved by the other three enzymes and should, therefore be read accordingly. For example, look at Figure 12.6. Wherever there is a band in the U2 lane (ie adenosine) there is also a band one fragment size lower in the S7 lane.

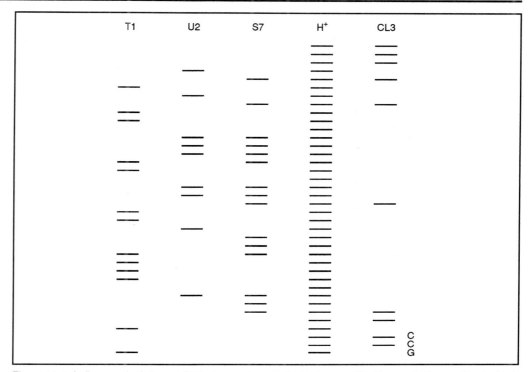

Figure 12.6 A diagram depicting an RNA sequencing gel. The headings (T1, U2 etc) designate the method of hydrolysis (see text for details). H⁺ represents acid hydrolysis.

The simplest way to read the sequences is to record the purines first and represent the other nucleotides as a dot. Then fill in the cytidines by reference to the CL3 lane and finally the uridines by looking at the bands in the S7 lane. Try it on SAQ 12.3. You will find the random hydrolysis (H⁺) lane very useful and leaving the uridine (S7) lane till last, less problematical. You could also try to obtain a sequence looking at the S7 lane first; you will find this much more difficult.

| SAQ 12.3 | Look at the diagram shown in Figure 12.6 representing a portion of an autoradiogram of an RNA sequencing gel, run following enzymatic cleavage of end-labelled RNA. What is the sequence, starting with G (in the T1 track) at the bottom of the figure? |

The other approach to RNA sequencing is to do primer extension reactions in the presence of dideoxynucleoside triphosphates. The principle is again exactly as for DNA; of course the enzyme reverse transcriptase has to be used, but this is often employed for DNA sequencing in any case, because it can use a DNA template and gives a more uniform incorporation of dideoxynucleotides in the growing chain than does the Klenow fragment of DNA polymerase. There are two difficulties with dideoxy RNA sequencing compared with DNA sequencing; in the first place a primer site needs to be known, because the RNA cannot be cloned, and secondly complementary chain sequencing cannot be carried out as for DNA, simply because RNA is single-stranded.

| SAQ 12.4 | 1) What methods are available for analysing: a) the base composition of RNA; b) the nucleotide sequence of RNA? |
| | 2) Suppose the RNA contains modified nucleotides, what do you think the effect will be upon RNA sequencing results? |

12.4 Structure of RNA

12.4.1 Secondary structure of RNA

So far we have only discussed RNA with reference to its primary structure. That is, we understand the basic structure of RNA to be a polynucleotide chain with 5'-3' sugar phosphate links. It is single-stranded and we have the means to determine the composition and sequence.

But do RNA molecules really exist simply as a single strand, rather like a length of string, randomly positioned in space; or do they form regions of secondary structure?

∏   What do you think the consequence would be if a region of nucleotides in RNA was followed by a complementary sequence further down the same molecule?

We can reasonably expect the chain to fold back on itself and form a base-paired structure. The complementary sequences are in close proximity; they are on the same molecule and, therefore, should be able to seek each other out in a hybridisation reaction.

hairpin   The resulting structure is referred to as a hairpin. It will have a stem, the double-stranded region, and a loop; that is a region of unpaired bases at one end (Figure 12.5). The size of the loop will depend on the distance between the complementary regions. It may also have small loops or bulges within the stem where regions do not properly pair up.

∏   This is fine in theory, but what evidence do we have that secondary structures exist in RNA? Try and think for a few moments of techniques that might illustrate the presence of secondary structure.

We have already discussed some. For example:

- the mobility of RNA changes during electrophoresis if we switch from native to denaturing conditions;

- sedimentation coefficients are not directly related to molecular mass;

- RNA exhibits hyperchromicity, in other words its absorbance of ultra-violet radiation increases as it is denatured; and

- nucleases that are specific for single-stranded sequences can be used to reveal double-stranded regions.

However, with the exception of the latter approach, these techniques do not show us which regions are double-stranded. The situation is complex because there could theoretically be many plausible secondary structures. However, if we know the primary structure we may be able to define many possibly overlapping regions that could base-pair. In which case how do we predict the most plausible structure? The answer is to apply rules that describe the interaction of base pairs. These rules can then be used to assess the relative stabilities of the range of possible structures and assess which is the most plausible.

This approach can, of course, only be an approximation to the true situation *in vivo* because we do not know the precise chemical and physical micro-environment of the RNA and it does not take into account interaction with other molecules such as proteins.

What are the rules? You may already have guessed; they involve thermodynamics, in fact a calculation of ΔG (the change in Gibbs function), for the formation of various structures. Thermodynamics are discussed elsewhere in this series (BIOTOL text, 'Principles of Cell Energetics', so to understand this section fully you may need to refer to that text). However, we can progress here with relatively little knowledge of thermodynamics. All we need to realise, or remember, is that reactions that require energy have a positive ΔG ($\Delta G > 0$) whereas those reactions which release energy have a negative value ($\Delta G < 0$); these latter reactions are spontaneous reactions.

Π Reactions with positive and negative ΔG exist; but what must the overall ΔG be for a series of reactions that occur spontaneously?

Common sense tells you it must be negative; experienced in everyday life as, 'You never get something for nothing!'

A revision note on thermodynamic principles

The free energy change (ΔG) of a reaction depends on two factors:

- the concentrations of reactants and products;
- the standard free energy change (ΔG°), ($= -RT \ln K$). This can be determined from tabulated values of the standard free energy of formation ($\Delta_f G^\circ$) of reactants and products. The standard state of a compound is the stable state in which the compound exists at 1 atmosphere pressure and at 298 K; by definition for elements in their standard states $\Delta_f G^\circ = 0$, in addition in biochemistry the standard state of the proton is chosen at a concentration of 10^{-7} mol l^{-1} (ie pH7) for this $\Delta_f G^\circ$ (H$^+$) = 0. (In chemistry the standard state of the proton is 1 mol l^{-1}, pH = 0). Using the biochemists' standard state, the symbol for the standard free energy is $\Delta G^{\circ'}$; K = equilibrium constant and is the ratio of the final concentrations (ideally activities) of products to the initial concentration of reactants.

Both these factors are taken into account in the Van't Hoff Isotherm, which for the general reaction in solution:

$$aA + bB + \text{---} \quad \rightarrow \quad lL + mM + \text{-.}$$

can be written:

$$\Delta G = \Delta G^{\circ'} + RT \ln \frac{C_L^l \, C_M^m}{C_A^a \, C_B^b} \qquad \text{(E - 12.1)}$$

where C_A, C_B -- are the initial concentrations of the reactants and C_L, C_M --- the final concentrations of products and a, b, l and m are the coefficients in the reaction equation.

It is obvious from Equation (E - 12.1) that ΔG values for a sequence of reactions are not additive, the $\Delta G^{\circ'}$ values, however, are additive.

The terms on the right-hand side of Equation (E - 12.1) determine together if ΔG will be negative or positive and thus both determine if the reaction will be spontaneous or not.

For the subject of folding and base-pairing in single stranded RNA molecules, the second term is negligible compared to the $\Delta G^{\circ'}$ value and so the free energy changes in the various pairings will be discussed in terms of the standard free energies.

Therefore, energy must be released to form a secondary structure, the overall free energy is given by:

$$\Delta G^{o'}_{total} = \Delta G^{o'}_i + \Delta G^{o'}_{sym} + \Sigma \Delta G^{o'}_x + \Sigma \Delta G^{o'}_u \qquad \text{(E - 12.1)}$$

$\Delta G^{o'}_i$ = change in free energy from initiation of a double helix. It is positive, $+ 14.2$ kJ mol^{-1} (3.4 kcal mol^{-1}) to form the first base pair.

$\Delta G^{o'}_{sym}$ = change in free energy for self-complementarity within a sequence. Think of it as the necessity for a molecule to bend or hybridise to itself. It is also positive, $+1.7$ kJ mol^{-1} (+0.4 kcal mol^{-1})

$\Sigma \Delta G^{o'}_x$ = the sum of the free energy charges for the individual base-pairing reactions shown in Table 12.4. These are negative.

$\Sigma \Delta G^{o'}_u$ = the sum of positive changes of free energy resulting from the close proximity of opposing bases that are not complementary. In a hairpin, the odd base may not match, the greater the number of these the greater the $\Sigma \Delta G^{o'}_u$ and the less favoured will be the overall structure.

When these base pairs form, they release free energy from two kinds of reaction; the formation of hydrogen bonds and base stacking.

Doublet	$\Delta G^{o'}$ (kJ mol^{-1} or kcal mol^{-1})
GC CG	-14.2 (-3.4)
GG CC	-12.1 (-2.9)
GA CU	-9.6 (-2.3)
GU CA	-8.8 (-2.1)
CG GC	-8.4 (-2.0)
CA GU	-7.5 (-1.8)
CU GA	-7.1 (-1.7)
UA AU	-4.6 (-1.1)
AA UU	-3.8 (-0.9)
AU UA	-3.8 (-0.9)

Table 12.4 Free energy of base pairing by nucleotide doublets.

∏   Look at the $\Delta G^{o'}$ figures in Table 12.4 and see if you can see a logic to them.

The simple point to come out is that G-C pairing releases more energy than A-U pairing, simply because of the increased number of hydrogen bonds. Far less obvious is the effect of base stacking. Stability is increased as water molecules are excluded from the centre of the helix and the free energy released depends on the particular combination of bases.

You should also note, at this stage, that G-U is a pairing that has a small negative $\Delta G^{o'}$ and is, therefore, well tolerated in hairpin structures.

To calculate the total free energy released by the formation of a duplex all of the individual doublet free energies need to be summed. Each base, except those at the ends is counted twice, as it is involved in two doublets, one for the base pair on either side. Also all the positive $\Delta G^{o'}$ values (ie $\Delta G_u^{o'}$) for internal loops and bulges (not given here) need to be summed and included in the equation.

Consider this simple example in Figure 12.7.

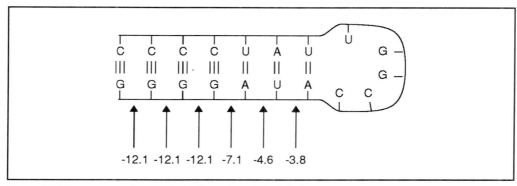

Figure 12.7 Free energy changes in the formation of a hairpin loop.

For a hairpin loop of 5 impaired bases like that shown in Figure 12.7, $\Delta G = +18.5 \, kJmol^{-1}$.

Using Equation (E - 12.1) and the data in Table 12.4 we can work our the free energy change for the formation of:

$$\Delta G_{total}^{o'} = \Delta G_i^{o'} + \Delta G_{sym}^{o'} + \Sigma\Delta G_x^{o'} + \Sigma\Delta G_u^{o'}$$

$$= 14.2 + 1.7 - 51.8 + 18.5 = -27.4 \, kJ \, mol^{-1}$$

Therefore, as $\Delta G_{total}^{o'}$ is negative, the above structure will form spontaneously from a single strand of RNA with the sequence CCCCUAUUGGCCAUAGGGG.

The end result is that we can predict the most thermodynamically stable secondary structures for RNA if we know the nucleotide sequence. This has been very valuable, for example in understanding tRNA structure and function. Other examples are the effect of secondary structure on the translation efficiency of mRNA, and the role of secondary structures in the control of gene expression, such as in the attenuation model for the control of the tryptophan operon in bacteria discussed elsewhere in the BIOTOL series (eg the BIOTOL text, 'Genome Management in Prokaryotes').

12.4.2 Tertiary structure of RNA

Do not be misled by the cloverleaf structure of tRNA (Figure 12.5). It is a useful model showing the regions of base-pairing within the molecule, but it is not a representation of the three-dimensional shape that tRNA adopts. Studies using X-ray crystallography show that tRNA has an L-shaped tertiary structure; the amino acid acceptor stem and TψC stem forming one double helix at right-angles to another helix formed by the D and anticodon stems (see Figure 12.5).

∏ What do you think holds the tertiary structure together?

The tertiary structure is stabilised by hydrogen bonding; largely involving bases not shown as paired in the clover leaf structure. Many of the bases commonly found in the same position in all tRNAs (the so-called invariant bases) or most tRNAs (the semi-invariant bases) are involved in this H-bonding. This provides an explanation for their conservation.

Not all tRNAs have been crystallised, but deductions leading to the common clover leaf structure suggest that all tRNAs have the same L-shaped tertiary structure (Figure 12.5). Note that the two functional components of the molecule, that is the anticodon and the amino acid binding site, are maximally separated. This similarity in shape presents a problem for the amino-acyl tRNA synthetases. How do they recognise the correct tRNA for addition of the amino acid? A current theory is that the enzymes make contact along the inside of the L-structure and that recognition involves fine subtlety in shape determined by perhaps just a few bases. These may be in the anticodon, theoretically perhaps the most obvious recognition point, such as for the tRNAs for valine and methionine. This has yet, however, to be proven.

12.5 Gene cloning and the study of RNA

In Chapter 11 the techniques available for isolating and purifying specific RNA were discussed. It might seem to be an essential prerequisite to studying RNA structure, but is it?

∏ What approach could you take to obtain the sequence of an RNA molecule, without ever purifying the particular RNA?

The answer is to use the range of gene cloning techniques now available. A gene can be cloned without needing to purify mRNA. The DNA can then be sequenced and the RNA sequence inferred. This procedure is discussed briefly in Chapters 6 and 11 and at more length in the BIOTOL text, 'Techniques for Transferring Genes'.

Once a gene has been cloned, by any of several strategies, it can be sequenced. Our knowledge of gene organisation is now such that the primary structure of RNA transcript can be deduced. For example, we know the general structure of promoters, (the DNA sequences responsible for the correct initiation of transcription) and can, therefore, identify the 5' end of a transcript; knowledge of the genetic code allows us to identify a coding region, the so-called open reading frame (ORF); and we can identify mRNA splice sites and transcription termination signals. What is more, we can alter the genes by a technique known as *in vitro* mutagenesis and re-introduce the DNA into a cell and observe effects on RNA and protein synthesis. This enables us to determine important control sites in DNA and features in RNA responsible for particular functions, such as mRNA - ribosome interactions in the initiation of protein synthesis.

Another approach, if bulk isolation and purification of a particular RNA is difficult (which it often is), is to simply make the RNA *in vitro* (Figure 8.8).

Cloned genes can be transcribed *in vitro* using a purified enzyme such as the SP6RNA polymerase (isolated from SP6 virus infected *Salmonella typhimurium*) or T7RNA polymerase (from phage infected *Escherichia coli*). Both of these enzymes are very active *in vitro* and can produce 10 µg of RNA from 1 µg of template in a simple reaction. The RNA product can then be used for any number of experiments where pure RNA is necessary; for example, the study of mRNA splicing where adequate amounts of substrate may otherwise be very difficult to obtain. Another use of these transcription systems is to produce labelled probes: if the reaction is carried out in the presence of labelled nucleotides, high specific activity RNA probes are synthesised that can be used in hybridisation reactions.

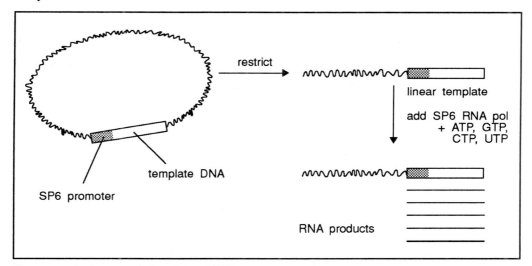

Figure 12.8 *In vitro* transcription of cloned DNA.

12.6 Catalytic activity

Enzymes are proteins: a dogma of biochemistry. But think of evolution. It is rather implausible to think of nucleic acids and proteins evolving simultaneously to create a replicating system. The answer to the paradox perhaps lies in a relatively recent discovery: the catalytic activity of RNA. This can be seen in ribonuclease P (a tRNA processing enzyme of *E. coli*) that contains two components: a 375 nucleotide RNA and a polypeptide of RMM 20 000. Given suitable conditions (determined by the Mg^{2+} concentration) the RNA component alone has very active catalytic activity. Indeed, the role of the protein appears to be to maintain the correct environment and therefore enhance the activity. Another example is that some RNA molecules are self-splicing, as illustrated in Figure 12.9.

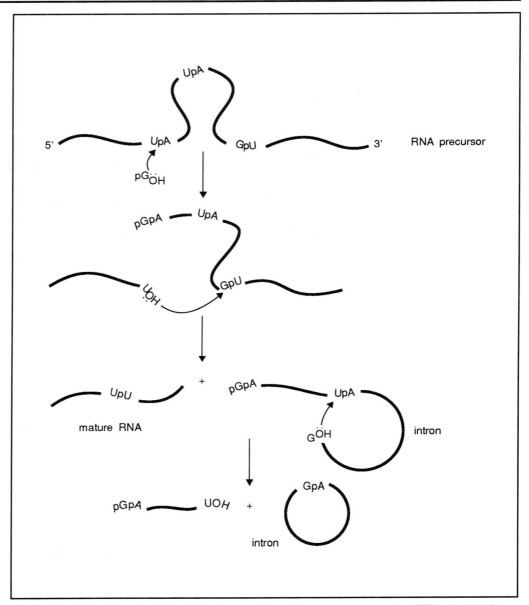

Figure 12.9 Self-splicing of a rRNA of *Tetrahymena*. No protein or energy source (eg ATP hydrolysis) is required. Note that guanidine first causes the RNA to hydrolyse at a uridine residue. The released RNA chain then catalyses the removal of a large nucleotide sequence. Subsequently the released RNA portion catalyses the removal of a further sequence by a process of autolysis (self hydrolysis).

Perhaps the primitive systems consisted of self-replicating RNA molecules with limited catalytic activity (we think it must be RNA because the 2'-OH is involved in splicing reactions), proteins and DNA evolving later. Certainly, catalytic activity of RNA resolves the 'chicken and egg' dilemma of splicing, ie what evolved first, the intron or the splicing enzyme.

Summary and objectives

The relative molecular mass of RNA can be determined by agarose gel electrophoresis under denaturing conditions. Base composition is determined by hydrolysis and chromatography alongside known standards. Modified nucleotides, such as exist in tRNA can be identified. RNA sequencing is carried out, in principle, as for DNA sequencing, viz: by enzyme or chemical cleavage of end-labelled RNA or by primer extension reactions with dideoxynucleosides, followed by acrylamide gel electrophoresis. RNA molecules have secondary structures that can be determined using thermodynamic principles. Tertiary structures have been studied using X-ray crystallography. Some cloning approaches have been developed that reduce the need to purify individual mRNA. Some RNA molecules have been shown to possess catalytic activity.

On completion of this chapter you should now be able to:

- select suitable methods for the determination of the relative molecular mass of RNA molecules;

- understand the approach to base-composition analysis;

- outline the methods used for RNA sequencing;

- describe the principles, but not the details, of how secondary and tertiary structures are determined;

- appreciate that DNA manipulation and cloning techniques have had a significant impact on studies on RNA structure and function.

Responses to SAQs

Responses to Chapter 1 SAQs

1.1: This was quite an open ended question. Here we hoped you would have been able to list a large number of techniques. We have divided the methods into physical and non-physical procedures.

Physical methods:

- solid shear methods - disruption of plant tissue, bacterial cells;

- liquid shear methods, blenders, homogenisers - disruption of plant tissues, bacterial cells;

- pressure homogenisation - disruption of cultured cells;

- tissue presses - disruption of bacterial cells;

- osmotic shock - disruption of animal cells;

- sonication - disruption of bacterial cells, animal cells.

Non-physical methods:

- chemical methods, eg organic solvents - disruption of animal cells;

- enzymic digestion - disruption of bacterial cells.

It is often difficult to assign a method absolutely to a given cell type. This is because combined treatment, eg the use of enzymic digestion, osmotic shock and detergent treatment are often used and also because, for example bacteria vary from fragile organisms which can be disrupted by digestive enzymes (eg lysozyme) to species with thick cell walls which can only be disrupted by vigorous mechanical means.

1.2 The factors to be considered in designing a suitable homogenisation medium include:

- the osmotic pressure, to prevent subcellular organelles from swelling or shrinking the medium is usually chosen to be either slightly hypo-osmotic or iso-osmotic to maintain morphological integrity;

- the pH, the medium is usually buffered to the physiological pH. For example for animal cells a pH of 7.4 is often used; a slightly more alkaline pH minimises the fragmentation of the plasma membrane;

- in some cases we may use chelating agents (eg EDTA) to remove divalent metal ions, such as magnesium and calcium, which are required by membrane proteases. These enzymes are therefore inhibited and nuclear integrity is maintained. On the other hand we may deliberately add divalent metal ions. For example, if we wish to prepare intact ribosomes, then the presence of magnesium ions is essential.

The main purpose of this question was to get you to realise that the design of a homogenisation medium very much depends on what we wish to use the tissue homogenate for.

1.3 The principal methods available for the separation of biologically important molecules from homogenates include:

Density differences

- centrifugation depends on the different densities of the species involved.

Solubility differences:

- liquid-liquid extraction depends on the different solubilities of the components in a two-component system of immiscible liquids; partition chromatography and reverse phase chromatography use this principle for separation of mixtures;

- precipitation is a crude separation technique in which the solubility of proteins is greatly reduced in the presence of large amounts of inorganic electrolytes.

Charge differences

- ion-exchange chromatography is based on the separation of ionic species on an insoluble matrix containing weak acidic or basic groups which can exchange with ions in the surrounding medium;

- zone electrophoresis is based on the separation of ionic species with different charges, in an applied electric field;

- isotachophoresis is based on the separation of species with different isoelectric points in a pH gradient under an applied electric field.

Size differences

- gel filtration chromatography separates molecules in a gel matrix, according to their shape and size in a gel matrix. Separation in gel electrophoresis may also be achieved by the sieving action of a gel.

1.4 An ideal gradient material should:

- be inert towards the biological material and the centrifuge tubes;

- not interfere with the monitoring of the sample;

- be easily separated from the material to be separated after centrifugation;

- be stable in solution and available in a pure form;

- exert a low osmotic pressure.

1.5   To produce a solution of linearly decreasing pH the diameters of the two chambers must be equal. Chamber B should be filled with a buffer solution of high pH (pH 9-10) and chamber A with a solution of dilute acid (for example, HCl). On mixing a solution of continuously decreasing pH will result.

1.6   We can use Equation (E - 1.2):

$$s = (\ln x_2/x_1)\,[\omega^2(t_2 - t_1)] \qquad\qquad (E - 1.2)$$

in either of two ways. In the first we can calculate the value of the sedimentation coefficient for successive time intervals:

$\omega = 56\,850 \times 2\pi/60 = 5953.32\ \mathrm{s}^{-1}$

$\omega^2 = 3.5442 \times 10^7\ \mathrm{s}^{-2}$.

Taking the readings in turn and substituting in Equation (E - 1.2):

$s = (\ln 5.55/5.50)/(3.5441 \times 10^7 \times 500) = 5.11 \times 10^{-13}\ s = 5.11\ \mathrm{S}$

$s = (\ln 5.60/5.50)/(3.5441 \times 10^7 \times 1000) = 5.08 \times 10^{-13}\ s = 5.08\ \mathrm{S}$

$s = (\ln 5.70/5.50)/(3.5441 \times 10^7 \times 2000) = 5.04 \times 10^{-13}\ s = 5.04\ \mathrm{S}$

$s = (\ln 5.80/5.50)/(3.5441 \times 10^7 \times 3000) = 4.99 \times 10^{-13}\ s = 4.99\ \mathrm{S}$

$s = (\ln 5.91/5.50)/(3.5441 \times 10^7 \times 4000) = 5.07 \times 10^{-13}\ s = 5.07\ \mathrm{S}$

$s = (\ln 6.01/5.50)/(3.5441 \times 10^7 \times 5000) = 5.00 \times 10^{-13}\ s = 5.00\ \mathrm{S}$.

The average value of the sedimentation coefficient is therefore 5.05 Sv. A more elegant way to obtain an average value from such a set of readings is to rearrange Equation (E - 1.2) to the form:

$$\ln\,(x(t)/x(0)) = \omega^2\,st$$

the plot of $\ln\,(x(t)/x(0))$ against t is linear and of slope $\omega^2 s$.

Draw up the following table:

time (s)	0	500	1000	2000	3000	4000	5000
x (cm)	5.50	5.55	5.60	5.70	5.80	5.91	6.01
$\ln (x(t)/x(0))$	0	0.0090	0.0180	0.0357	0.0531	0.0719	0.0887

The graph shown on the next page has a slope of $1.789 \times 10^5\ \mathrm{s}^{-1}$.

$\therefore\quad \omega^2 s = 1.789 \times 10^5$ and

$s = 1.789 \times 10^5/3.5442 \times 10^7$

$= 5.048 \times 10^{-13} s = 5.048\ \mathrm{S}$.

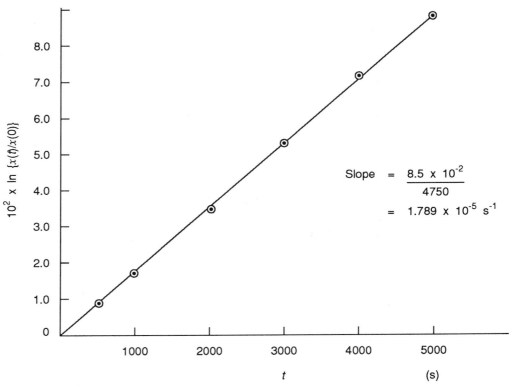

1.7   Plot to determine sedimentation coefficient.

1)   False. The force on a particle in an applied centrifugal field depends on its distance from the centre of rotations and on the angular velocity:

The applied centrifugal force = $\omega^2 x$, where ω is the angular velocity of the rotor and x the radial distance of the particle from the centre of the rotor.

2)   False. Sedimentation of a heterogeneous suspension does not result in a sediment which is homogeneous; it will contain particles of smaller mass that were situated near the bottom of the tube before centrifugation started and particles of greater mass that were originally located further up the tube.

3)   True. The rate of sedimentation of a macromolecule depends on both its shape and molar mass.

4)   False. In isopycnic centrifugation particles separate according to their buoyant density to the region where their density is equal to that of the medium.

1.8   Adsorption occurs through such intermolecular interactions as 2) dipole-dipole interaction, 3) van der Waals forces and 4) hydrogen bonds and ionic interactions (1) play a part especially in ion exchange chromatography.

1.9   Your completed table should look like this; if not then it is suggested you revise this important piece of work.

Type of chrom-atography	Polarity of stationary phase	Polarity range of mobile phase	Order of elution polar/non-polar	Effect of increase of polarity of mobile phase
normal phase	polar	weak to medium polarity	most polar eluted last	retention times decrease
reverse phase	non-polar	strong to medium polarity	most polar eluted first	retention times increase

1.10   You should have realised that below their isoelectric point (pI) proteins are positively charged and they are negatively charged above. Thus the proteins will be eluted sequentially as the pH of the eluting medium reaches the individual isoelectric points (ie the pH at which the protein is uncharged) of the proteins.

1.11   Your completed table should look like this:

Name of technique	Nature of stationary phase	Mobile phase	Mechanism of separation	Applications, separation of:
gas liquid chromatography	polyethylene glycols on inert support	inert gas	partition gas/liquid	volatile esters, hydrocarbons
adsorption chromatography	silica, alumina	polar solvent	adsorption	leaf and plant pigments
reverse phase chromatography	silica with bonded alkyl-silyl groups	water with methanol, ethanol or acetonitrile	hydrophobic interaction	peptides, nucleic acids
partition chromatography	water bound in inert matrix eg paper	organic solvent	partition liquid/liquid	amino acids, sugars etc
ion exchange chromatography	matrix carrying ionisable groups	buffer solution with pH or salt gradient	charge	proteins amino acids etc
gel filtration chromatography	cross-linked gel (eg agarose)	buffer solution or organic solvent	size	proteins nucleic acids
affinity chromatography	matrix carrying specific binding agent (eg antibody)	buffer solution	biological affinity	antigens

1.12   In Figure 1.9 the mixture of proteins A + B is applied at pH6 (the hatched area) in a linear pH gradient. At pH 6 protein A (pI = 5) will be negatively charged and so will move towards the positive electrode at an ever decreasing velocity until it reaches its isoelectric point where it focuses. Protein B, initially at a pH below its isoelectric point (pI = B), will be positively charged and so will migrate towards the negative electrode until it focuses at its isoelectric point. Note that the point of application of the sample is immaterial, each protein will focus at its own pI.

1.13 1) False; if you combine Equation (E - 1.6) and (E - 1.8) you obtain the relationship:

$$E = hc/\lambda$$

from which it is apparent that E is inversely proportional to the wavelength.

2) True, see Equation (E - 1.8).

3) False, since UV radiation is of short wavelength and infrared of longer wavelength, it follows from Equation (E - 1.5) that UV radiation has the greater frequency.

4) True, this follows directly from the previous statement and Equation (E - 1.8).

1.14 The answer is 3), ie electronic transitions from the ground or low energy levels to high energy levels. If you chose 4) you were describing fluorescence or phosphorescence; that is the energy released when electrons fall back to a lower energy level it may be emitted as light. Answers 1) and 2) are more appropriate to infrared spectroscopy. If you are unsure of this topic then you are strongly advised to revise it in a standard text-book.

Responses to Chapter 2 SAQs

2.1   If we assume that a 1 mg cm^{-3} solution of protein has an absorbance of 1, then it follows that the diluted solution ($A = 1.3$) has a concentration of approximately 1.3 mg cm^{-3}. The original solution was diluted 0.1 cm^3 in 3cm^3, ie 30 fold. Thus the approximate concentration of the original solution is 30 x 1.3 = 39.0 mg cm^{-3}.

Remember that rarely do we need a highly accurate measure of the amount of protein present (except of course when we have the final purified protein) and a quick and easy method such as this to obtain an approximate value is very useful.

2.2   Using equation (E - 2.1) $A = \varepsilon c l$

$0.25 = \varepsilon \times 0.4 \times 1$

$\therefore \varepsilon = 0.625 \text{ cm}^2 \text{ mg}^{-1}$.

For the second sample you can now use the extinction coefficient that you have just calculated;

$\therefore 0.08 = 0.625 \times c \times 1$

$c = 0.08/0.625$

$= 0.128 \text{ mg cm}^{-3}$

Therefore the mass of protein in 20 cm^3 is 20 x 0.128 = 2.56 mg.

2.3   The absorbance at 220 nm is measuring the peptide bonds present and is therefore a reasonable measure of the amounts of each protein present. There is therefore about five times as much protein A as protein B. The absorbance at 280 nm is only measuring the tyrosine and/or tryptophan content of each protein. Obviously protein B has a considerably higher tryptophan/tyrosine content than protein A and hence the total absorbance at 280 nm is approximately equal to that of protein B, even though there is far less of protein B than protein A.

Peak C is obviously due to a peptide or protein that does not contain either tyrosine or tryptophan. It cannot therefore be monitored at 280 nm but can be detected by the absorption of the peptide bonds at 220 nm.

2.4   From the calibration graph (shown on the next page) you should have determined that the BSA concentration in the unknown sample was 1.02 mg cm^{-3}. Since colour development in the Bradford assay depends on the presence of certain amino acid sidechains, it is not possible to use a BSA standard curve to determine the absolute concentration of any other protein. In other words, do not be tempted to deduce from the graph that the concentration of protein B is exactly 0.52 mg cm^{-3}.

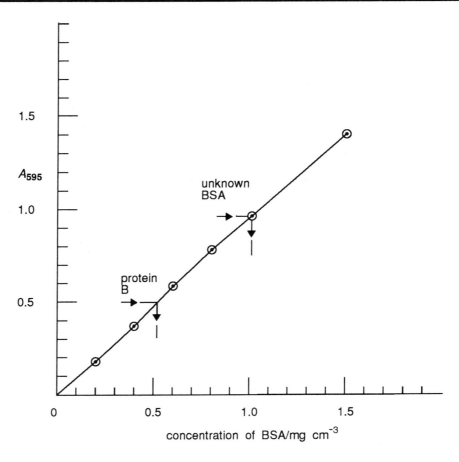

Calibration curve of A_{595} of against concentration for a BSA solution.

It is, however, probably a reasonable approximation of the concentration of protein B. Only if identical concentrations of BSA and protein B gave the same colour intensity with the Bradford reagent could you be justified in using the same calibration curve to obtain an accurate measure of the concentration of protein B. However, it is unlikely that two different proteins will give identical colour intensities with this test. The only real way to get an accurate quantitative assessment of protein B with this test is to prepare an initial calibration curve using known amounts of pure protein B and then relate all the results to this graph.

Unfortunately one often sees workers preparing a BSA standard curve and then using this to calculate a supposedly accurate concentration of another protein. Although this will give a reasonable approximation of the protein content, it will almost certainly not be a highly accurate figure.

2.5   Coomassie brilliant blue was the colour reagent used in the Bradford method for protein estimations; it is also used to stain protein on polyacrylamide gels.

2.6   Methods for detecting the presence of protein in solution include:

Measurement of absorbance at 220 nm; measurement of absorbance at 280 nm; the Biuret assay; the Lowry assay; the Bradford assay; amino acid analysis; gel electrophoresis.

2.7   The two amino acids responsible for absorption at 280 nm are tyrosine and tryptophan.

2.8   1) True, the higher the concentration of acrylamide the smaller the pore size.

2) False, the method separates proteins according to size.

3) False, a pure protein consisting of two unequal subunits will show two bands (one for each subunit) on the gel.

4) False. The Bradford method is a spectrophotometric method for quantifying proteins. It does, however, involve the use of the same dye (Coomassie brilliant blue) that is used to stain proteins on polyacrylamide gels.

5) False. Absorbance at 220 nm, the Bradford method and amino acid analysis are all sensitive below this level.

6) False. They are separated according to their charge.

7) False. It will have an absorbance of approximately 1.

2.9   Non-destructive method - measurement of absorbance at 220 nm or 280 nm.

Destructive method - Biuret, Lowry, Bradford and amino acid analysis.

Responses for Chapter 3 SAQs

3.1 For amino acids to be negatively charged at pH 8.5 means that they must have a pI > 8.5. These include lysine, arginine and ornithine.

3.2 1) False, generally the converse is true, therefore 4 is correct.

2) False, although many hydrophobic groups are to be found inside proteins, some hydrophobic groups are found on the surface.

3) False, the overall charge on the surface of a protein can be positive, negative or zero depending on the pH.

4) This is correct.

5) Incorrect, this statement is only true when the protein is in a solution where the pH of the solution is equal to the isoelectric point of the protein;

6) This is correct.

3.3 Below its isoelectric point a protein becomes progressively more positively charged (refer back to Figure 3.2 is you are unsure about this). Therefore, at pH 3.5 ovalbumin will have an overall positive charge. Conversely, above the pI value, a protein has an increasingly negative charge; therefore at pH 8.6 ovalbumin will have an overall negative charge.

3.4 For enzyme A: original specific activity = $45/50 = 0.9$ units mg^{-1}.

After heat treatment, specific activity = $2/15 = 0.13$ units mg^{-1}. Since specific activity should increase at each step this is obviously not a useful method as the specific activity has gone down from 0.9 to 0.13. Also the yield of enzyme at this step is $2/45 \times 100 = 4.5\%$. A yield as low as this is totally unacceptable.

For enzyme B: original specific activity = $120/50 = 2.4$ units mg^{-1}.

After heat treatment, specific activity = $60/15 = 4.0$ units mg^{-1}. This treatment has obviously resulted in some purification since the specific activity has gone up from 2.4 to 4.0 (a $4.0/2.4 = 1.6$-fold purification). However, the yield of enzyme is only 50% ($60/120 \times 100$). A loss of 50% of your enzyme in a single step is normally unacceptable (perhaps except when the step results in a dramatic increase in protein purification) and therefore this would not be a useful method for purifying enzyme B.

For enzyme C: original specific activity = $85/50 = 1.7$ mg^{-1}.

After heat treatment specific activity = $83/15 = 5.5$ mg^{-1}. We can see that the specific activity has increased from 1.7 to 5.5 (a 3.2 fold purification) and also the yield ($83/85 \times 100 = 97\%$) is extremely good. This is therefore a good purification step for enzyme C.

3.5   1) For $NaNO_3$:

$$I = \frac{1}{2}[c(Na^+)z^2(Na^+) + c(NO_3^-)z^2(NO_3^-)]$$

$$= \frac{1}{2}[0.5 \times 1^2 + 0.5 \times 1^2] = 0.5 \text{ mol } l^{-1}$$

2) For Na_2SO_4:

$$I = \frac{1}{2}[c(Na^+)z^2(Na^+) + c(SO_4^{2-})z^2(SO_4^{2-})]$$

$$= \frac{1}{2}[2 \times 0.5 \times 1^2 + 0.5 \times 2^2] = 1.5 \text{ mol } l^{-1}$$

3) For $(NH_4)_2SO_4$:

$$I = \frac{1}{2}[2 \times 1.5 \times 1^2 + 1.5 \times 2^2] = 4.5 \text{ mol } l^{-1}$$

4) For mixture of $NaNO_3 + (NH_4)_2SO_4$, we have to take all this into account:

$$I = \frac{1}{2}[c(Na^+)z^2(Na^+) + c(NO_3^-)z^2(NO_3^-) + c(NH_4^+)z^2(NH_4^+) + c(SO_4^{2-})z^2(SO_4^{2-})]$$

$$= \frac{1}{2}[0.5 \times 1^2 + 0.5 \times 1^2 + 2 \times 2.0 \times 1^2 + 2.0 \times 2^2]$$

$$= \frac{1}{2}[0.5 + 0.5 + 4.0 + 8.0]$$

$$= \frac{1}{2}[13.0] = 6.5 \text{ mol } l^{-1}$$

From this you have probably noticed two important points:

- $I = c$ for 1,1 electrolytes;

- the importance of the charge, there being a very marked increase in I for electrolytes containing polyvalent ions.

3.6   If you have decided to collect the fraction that precipitates at 55% saturation you obviously understand what you are doing. By adjusting the extract initially to 45% ammonium sulphate and centrifuging we will remove 650 mg of protein, but only lose 15 units (~5%) of our enzyme. If we now make the supernatant 55% saturated in ammonium sulphate, most of our enzyme will be in the precipitate that forms, whereas 550 mg of protein will still remain in solution (as well as 25 units of enzyme). We should therefore collect the 55% precipitate and re-dissolve this in buffer solution. The original specific activity of the enzyme was $310/1700 = 0.18 \text{ mg}^{-1}$. The specific activity of the 55% ammonium sulphate fraction is $270/500 = 0.54 \text{ mg}^{-1}$. We therefore have a threefold purification ($0.54/0.18 = 3.0$) and a yield of 87% ($270/310 \times 100$) is quite acceptable for this method.

3.7   We would expect five peaks of protein. The first five proteins in this list have molecular masses greater than the exclusion limit (30 000) of G50 and therefore will all elute together in the totally excluded (or void) volume. The molecular masses of the remaining four proteins fall within the fractionation range of G50 (30 000–1500) and therefore will elute from the column in order of decreasing molecular mass, ie trypsin will elute first, followed by soya bean trypsin inhibitor, lysozyme and finally insulin.

Would a better separation have been achieved using a G-75 column?

Yes it would. Phosphorylase and transferrin would still elute in the totally excluded volume, but the remaining seven proteins, being within the exclusion limit (70 000) of G-75 should separate. However, note that because of the similar molecular masses of ovalbumin and hexokinase, it is unlikely that these two proteins would separate and would almost certainly elute in the same position.

3.8   Your aim here is to find a pH value at which your protein of interest has the opposite charge to all the others. You then apply your mixture at this pH to an ion exchange column that has the same charge as your protein of interest. This protein will pass straight through the column, whereas the others, having an opposite charge to that on the resin, should remain in the column. For example, at say pH 8.0, lysozyme will have an overall positive charge being below its p*I*, whereas the other proteins will all have negative charges being above their p*I* values. If the mixture is therefore applied at pH 8.0 to a DEAE-cellulose column, the lysozyme will pass through the column unbound, whereas the other proteins will bind.

Similarly at pH 4.0, fetuin will have a negative charge and will not bind to a CM-cellulose column, whereas the other proteins will have an overall positive charge and will be retained on the column.

3.9   2); 5); 7) and 8) are not used to purify proteins, the remaining methods are all suitable.

3.10   1) and 4) are correct. Proteins can be eluted from the exchange columns by either a change of pH or by applying a salt gradient.

3.11   1) and 4) are correct. Ammonium sulphate precipitation is used because the material is cheap and the process can be carried out on a large scale. It is non-specific and can be used for separation of any proteins, including enzymes.

3.12   Affinity chromatography is very specific depending on a support ligand which has a shape and chain complimentary to that on a region of the protein. Hence 4) is correct.

3.13   The isoelectric point of a protein (or an amino acid) is the pH at which the numbers of positive and negative charges are equal and the overall charge on the protein is zero; hence 4) is correct.

Responses to Chapter 4 SAQs

4.1   It is first necessary to calculate the volume of each bead and then the total number of beads in the 10cm^3 volume. The area of the beads is also calculated

volume of sphere = $\frac{4}{3} \pi r^3$

area of sphere = $4 \pi r^2$.

	0.2 mm bead	5 μm bead
volume of bead	$\frac{4}{3} \pi \left(0.1 \times 10^{-3} m \right)^3$	$\frac{4}{3} \pi \left(2.5 \times 10^{-6} m \right)^3$
	$= 4.19 \times 10^{-12}$ m^3	$= 6.55 \times 10^{-17}$ m^3
no of beads in 10 cm^3	$\dfrac{\left(10 \times 10^{-6} m^3 \right)}{\left(4.19 \times 10^{-12} m^3 \right)}$	$\dfrac{\left(10 \times 10^{-6} m^3 \right)}{\left(6.55 \times 10^{-17} m^3 \right)}$
	$= 2.39 \times 10^6$	$= 1.53 \times 10^{11}$
total area of beads	$= 4 \times \pi \times \left(0.1 \times 10^{-3} m \right)^2 \times 2.39 \times 10^6$	$= 4 \pi \left(2.5 \times 10^{-6} m \right)^2 \times 1.53 \times 10^{11}$
	$= 0.30$ m^2 (the area of a small coffee table)	$= 12.0$ m^2 (the area of a small room)

Thus for the given volume the surface area exposed by the smaller beads is 40 times that provided by the larger beads. This is a somewhat idealised calculation since it assumes that the beads are ideally packed with no spaces between them. In practice this could only be achieved if the beads were cuboid. Nevertheless, we have made the point that the smaller the bead, the greater the surface to volume ratio.

4.2   Peptide A has alanine at the N-terminus, and since it was produced by tryptic cleavage the arginine residue must be at the C-terminus. The sequence must therefore be Ala-Glu-Arg. By a similar argument the sequence of peptide C must be Asp-Glu-Lys. Peptide B has N-terminal phenylalanine and therefore must be the dipeptide Phe-Ser. Now we must decide in which order to place these peptides. We already know that the original peptide has N-terminal aspartic acid, so peptide C must be the N-terminal sequence. The carboxypeptidase data also shows that the C-terminal amino acid is serine. So peptide B must be at the C-terminus.

Peptide A must therefore fit in the middle giving the sequence:

Asp-Glu-Lys-Ala-Glu-Arg-Phe-Ser.

There was another reason why we could deduce peptide B was the C-terminal peptide. Following a tryptic digest all peptides produced will end in arginine or lysine except the peptide from the C-terminus. Any peptide not ending in arginine or lysine must come

4.3   1) False, gramicidin is an antibiotic.

2) False, peptides are present in extremely low concentrations and remain in solution on the addition of ammonium sulphate.

3) True, peptides are smaller than proteins.

4) True, PTH amino acid can be analysed by reverse phase chromatography.

4.4   You have been sequencing a mixture of two peptides. The two peptides obviously have similar characteristics and were eluted together from the reverse phase column. They both obviously have N-terminal valine which led you to believe you had a pure peptide. However, once you cleave off this first N-terminal amino acid, it is obvious you are now sequencing two peptides.

Responses to Chapter 5 SAQs

5.1   The following criteria are commonly used to characterise a protein:

- biological activity (for example, does it have enzymatic activity?);

- molecular mass (and, where relevant, number of subunits);

- amino acid analysis;

- N-terminal sequence.

You should realise that additional criteria could also be used. For example, we could use the number and sizes of peptides produced after hydrolysis by a particular proteinase.

5.2   Your graph should be like that shown in Figure SAQ 5.2. This gives a RMM of aspartate aminotransferase of approximately 40 000. The value for β-lactoglobulin is not an error and will be discussed later.

(NB The relationship between the distance moved in the gel and RMM is purely empirical).

Protein	RMM	log RMM	Distance moved (mm)
transferrin	78 000	4.89	13
bovine serum albumin	66 000	4.82	21
ovalbumin	45 000	4.65	35
β–lactoglobulin	36 800	4.57	77
carbonic anhydrase	30 000	4.48	56
trypsinogen	24 000	4.38	64
myoglobin	17 800	4.25	80
cytochrome C	12 400	4.09	95
AAT			42

Notice the apparent discrepency of the distance moved and RMM value for β-lactoglobulin. We will return to this in the main body of text.

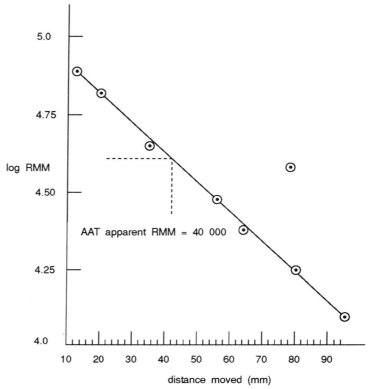

SAQ 5.2 Graph of log RMM against distance moved in gel electrophoresis.

therefore for AAT log RMM = 4.6 ∴ RMM = 40 0000.

5.3 Your plot of log RMM against elution volume should be like that shown in Figure SAQ 5.3. Thus from the gel filtration data the RMM is about 80 000. This is double the answer obtained by SDS gel electrophoresis. The reason for this is that AAT is a dimer of two equal subunits of RMM 40 000. On gel filtration one is studying the native enzyme (ie the two subunits are linked to give a protein of RMM 80 000) whereas in SDS gel electrophoresis the denaturing conditions disrupt the native protein and we are observing just the separate subunit chains.

Protein	RMM	log RMM	Elution volume (cm^3)
phosphorylase B	97 400	4.99	102
transferrin	78 000	4.89	121
bovine serum albumin	66 000	4.82	130
ovalbumin	45 000	4.65	154
β-lactoglobulin	36 800	4.57	170
α-chymotrypsin	22 500	4.35	197
lysozyme	14 300	4.16	235

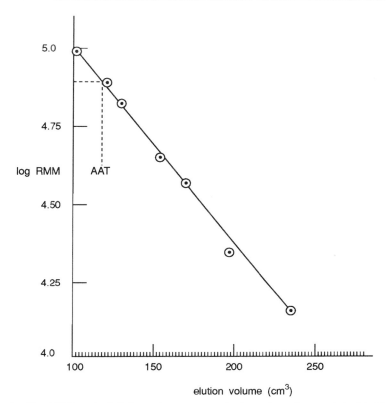

SAQ 5.3 Graph of log RMM against elution volume on a sephadex column for a series of proteins.

for AAT; $\log \text{RMM} = 4.88$ ∴ $\text{RMM} = 78\,000$.

5.4 Since haemoglobin consists of two α- and two β-chains, each sample should show two bands on the gel, one corresponding to the α-chain, the other the β-chain. Since SDS gels separate according to size, the difference in the β-chains of HbA and HbS are negligible as far as size is concerned and thus the two bands from each sample will run in exactly the same position. Thus we would be unable to detect the differences between HbA and HbS.

5.5 It is fairly obvious that the peptide produced by endoproteinase Glu-C was produced by cleavage after the Asp residue in peptide 1 and after the Glu residue in peptide 3. We can therefore link peptide 1 to 3 to give the sequence: Ser-Ala-Val-Leu-Gly-Asp-His-Phe-Arg-Ala-Pro-Ser-Glu-Val-Thr-Gly-Phe-Leu-Ileu-Arg.

5.6 Trypsin cleaves C-terminal to arginine and lysine residues. It would therefore cleave the peptide into four pieces, cleaving after residues 3, 8 and 12. This would not give us what we want.

Endoproteinase Glu-C will cleave after residues 6 and 11 giving three peptides. One of these will contain the first six residues of the peptides and may be useful for studying the function of the N-terminal sequence. However, the C-terminal sequence will be cleaved into two pieces by cleavage after residue 11.

Chymotrypsin should cleave only after the phenylalanine residue at position 7. This will give two seven residue peptides, one corresponding to the N-terminus of the peptide, the other to the C-terminus.

Responses to Chapter 6 SAQs

6.1   The most obvious difference between RNA and DNA is reflected in the names of these molecules: RNA contains ribose in its 'backbone' but DNA contains deoxyribose. The other difference concerns the nucleotide bases: thymine is normally found only in DNA and uracil only in RNA; the other bases (cytosine, adenine and guanine) occur in both RNA and DNA.

6.2   1) The bases in RNA are: guanine, G; cytosine, C; adenine, A; uracil, U; and the bases in DNA are: guanine, G; cytosine, C; adenine, A; thymine, T.

2) DNA lacks a hydroxyl on the 2' carbon of the ribose, it is replaced with a single H atom; DNA is usually double stranded; DNA is generally of higher molecular mass (10^6 g); RNA about 25 000 g.

3) The base-pairing rules are: A=T; A=U; G≡C.

Since RNA molecules consist of a single strand in the form of a random coil the base pairing relationships do not apply.

6.3   1) This experiment indicates that two of the fragments of bacterial DNA separated by electrophoresis contained nucleotide sequences that were complementary to nucleotide sequences present in the bacteriophage. It could indicate (but not prove) that the original bacteria were 'contaminated' by bacteriophage.

2) This experiment would indicate that the bacterium carried a gene which coded for β-galactosidase. Thus, the ^{32}P-mRNA would hybridise with this sequence. This in turn would cause a blackening of the film. We could use the position of this blackened area on the film to locate the DNA fragment carrying the β-galactosidase gene on the original electrophoretogram.

6.4   1) Nucleotide fragments a) and c) will anneal since the overlapping ends are complementary. We can draw this as:

a)

Sequences b) and d) are not entirely complementary and therefore less likely to anneal.

2) A genome library is **more** likely to contain a particular gene since it potentially contains **all** of the nucleotide sequences (and thus all genes) present in the genome. The cDNA library will only contain nucleotide sequences that are expressed. Of course, if we know that the gene of interest is expressed, using a cDNA library rather than a genome library is more likely to lead to the successful isolation of the gene. This is because nucleotide sequences of the gene will be in the form of mRNA.

Conversion of this in RNA into cDNA means that we will have amongst the various cDNA molecules intact sequences corresponding to the gene of interest.

3) The answer to this is implicit in the response given to 2). To appear in a cDNA library a gene has to be expressed and its transcript (RNA) isolated. If the gene is repressed then no RNA will be produced and thus the corresponding cDNA will not be present in the cDNA library. Likewise nucleotide sequences which perform a structural function(eg as physical spacers) and are never expressed as RNA will not appear in the cDNA library.

Responses for Chapter 7 SAQs

7.1
1) Liver - a good choice since it is soft and readily homogenised.

2) Bone - almost impossible to break up and contains few cells, so not a good starting point.

3) Red blood cells - appear attractive since they are easily broken open. However, mature red blood cells do not contain nuclei and so would be of no use to us.

4) Leaf buds - a rich source of nuclear DNA. Their cells have not yet expanded and so the buds contain a large number of small cells per unit volume. The tissue is relatively soft and so is easily homogenised.

5) Potato tubers - not easily homogenised unless cooked! This is also a poor choice because its cells are relatively few in number per unit volume, having expanded greatly to act as stores of starch.

7.2
The main steps required to prepare an extract of DNA from peas are:

1) harvest young, but well greened pea leaves. These should contain a large number of chloroplasts. You would not be wise to use pea seeds or roots since these will contain few chloroplasts;

2) homogenise the leaves gently using a blender or pestle and mortar in the presence of a buffer solution containing EDTA. Do this on ice to help keep the organelle membranes stiff and so increase the yield of intact chloroplasts;

3) filter the extract to remove cell debris and then collect chloroplasts by differential centrifugation;

4) lyse the chloroplasts using Triton X-100. This will not lyse any nuclei which may still be contaminating the chloroplasts;

5) centrifuge to sediment any intact nuclei and recover the supernatant which should contain chloroplast DNA in solution.

This is as far as you need to go for this exercise. The DNA will not, of course be free of RNA or protein, so it will need further steps to purify it completely.

7.3
1) This extract from pea leaves would have DNA, RNA and many other types of molecule in it, including chlorophyll. Absorbance at 260 nm would be useless, not only would RNA contribute to the absorbance but so would chlorophyll. Since the DNA is likely to be present at a fairly high concentration the Burton assay would be appropriate. The sensitivity of the DABA assay would not be needed and chlorophyll might interfere with the fluorescence. A small sample could be run on a gel and stained with ethidium bromide but, since this is a crude extract, the presence of proteins, lipids, etc might interfere with the running of the gel.

2) This sample of fairly concentrated DNA is ideally suited to assay by absorbance at 260 nm. There is sufficient sample to fill a cuvette and none of the sample is consumed by this assay. No contaminants are present, so only DNA will absorb UV radiation. All the other assays could be used but they are more time-consuming and no more accurate.

3) This small sample extracted from a blood stain contains only DNA and RNA. It is typical of samples obtained for forensic investigations and, presumably, could be used for DNA 'fingerprinting' or 'profiling'. It is important not to destroy a significant proportion of the sample but extreme accuracy would not be needed. Such a sample is well suited to analysis by gel electrophoresis. Only a small volume would be loaded on the gel and DNA could be distinguished from RNA. The method has the added advantage that it will indicate whether the DNA is reasonably intact and so useful for further analysis.

Responses to Chapter 8 SAQs

8.1

1) Not true. The electric field is needed to cause movement of the DNA but it is not responsible for separation according to size.

2) Not true. Since the charge on a DNA molecule is generated by its phosphate groups, there will be a constant charge per unit length of the DNA.

3) True. The pores of the gel present an obstacle to the movement of DNA molecules, the resistance to movement becoming greater the larger the molecule.

4) Not true. The negatively charged DNA is attracted to the positive anode, but this is not the basis of size discrimination.

5) Not true. The pH influences the charge on the DNA but all molecules in the same buffer gel will have the same charge per unit length. This is not, therefore, a factor in separating DNA molecules according to size.

8.2

You should have listed the following items needed for the recovery of a band of DNA from an agarose gel:

- ultraviolet transilluminator for viewing bands of DNA stained with ethidium bromide;

- perspex screen or mask for protection from UV radiation;

- scalpel for cutting out a slice of gel containing the desired band;

- dialysis tubing into which the gel slice is placed (do not forget the forceps for handling the slice);

- gel running buffer solution to go within and around the dialysis tubing;

- electrophoresis tank or similar container fitted with electrodes; the sealed dialysis bag is placed in this, under buffer solution;

- a power supply of the sort used for electrophoresis;

- scissors to cut open the dialysis bag;

- Pasteur pipette or similar to draw buffer solution containing eluted DNA out of the dialysis bag;

- disposable gloves to be worn at all times when handling solutions containing DNA.

Your list would probably end here, though you might have continued to include a phenol/chloroform extraction and ethanol precipitation to clean up and concentrate your sample.

8.3

1) Cutting plasmid DNA will generate linear fragments. Unless there is an extremely unusual distribution of bases within the plasmid the buoyant densities of the two fragments should be virtually identical. Both fragments will be free to unwind to allow the maximum possible binding of ethidium bromide. Thus both fragments will behave identically in the absence and in the presence of ethidium bromide, generating a single band in both cases.

2) In the absence of ethidium bromide the two forms of plasmid will have identical buoyant densities; hence a single band will be observed. Saturating levels of ethidium bromide will result in more dye binding per unit length to the open circle than the supercoiled DNA. Therefore the buoyant density of the open circle will be decreased to a greater extent than that of the supercoil. Two bands will be seen, the lower containing supercoiled, the upper open circle DNA.

3) All three DNA species have virtually the same density in the absence of ethidium bromide so only one band will be seen (a difference in density of 0.001 g cm^{-3} will not be resolved by this method). With ethidium bromide the linear and open circle species should take up equal amounts of dye per unit length, but the supercoil will take up less. Consequently there will be two bands of DNA: the lower will contain supercoiled plasmid, the upper will contain both open circle plasmid and linear chromosomal DNA.

Responses to Chapter 9 SAQs

9.1 Which of the three bands did you use for this estimation? You should have based your calculation on the middle band, since this is the linear form of the plasmid. The marker molecules are linear and so the calibration curve can only be applied to other molecules that are linear. The actual size of pBR322 is 4.36 kb.

9.2 One band in sample A must be single-stranded since it disappears on treatment with S1 nuclease; the other band and that in sample B must be double-stranded. All the DNA in sample A must be linear since both bands are removed by exonuclease treatment. Sample B DNA survives both S1 nuclease and exonuclease and so must be a covalently closed, double-stranded circle.

It is not possible to say if sample B contains open circle or supercoiled DNA. Treatment with topoisomerase would resolve this question. If the band of DNA remained unchanged we could conclude that it was in the open circular, relaxed conformation; a shift to one or more slower moving bands would indicate a degree of supercoiling in the DNA.

9.3 The percentage of adenosine bases in a DNA sample whose T_m was found to be 85.4°C can be obtained by using Equation (E - 9.1), relating GC content to T_m:

$$85.4 = (39.7 \times GC) + 69.0$$

$$\therefore \; 39.7 \times GC = 85.4 - 69.0 = 16.4$$

$$\therefore \; GC = 16.4/39.7 = 0.41$$

$$\therefore \; AT = 0.59$$

$$\therefore \; A = 0.59/2 = 0.30$$

$$\therefore \; \text{adenosine content} = 30\%.$$

To carry out this calculation, you hve to assume that the ionic conditions in which the T_m was determined were the same as those which were used in establishing Equation (E - 9.1)

9.4 The value of C_ot at which the DNA is 50% renatured is approximately 10^{-2} mol l^{-1}s nucleotides for organism A and 10 mol l^{-1}s nucleotides for organism B. The $(C_ot)_{0.5}$ value is clearly much higher for organism B, so this has the higher complexity and the larger genome.

9.5 In order to read a sequence from Figure 9.6 you must identify the shortest DNA chain. This will have moved farthest along the gel, so will be at the bottom of the autoradiograph. This is clearly in track 'C', so the first nucleotide in the sequence is C. The next chain in size is found in the 'A' track, so the next nucleotide in the newly synthesised chain is A. Continuing in this fashion you should be able to build up the following sequence:

CAGTTAGCCATTTTCACATGGTTACAGTAACTGTC

From about this point the bands are so compressed that it is impossible to be certain of their order. To read further along the sequence it would be necessary to run another gel for considerably longer in order to resolve these larger molecules.

Responses to Chapter 10 SAQs

10.1

1) Protein may be removed by denaturing agents (eg phenol, guanidinium salts) or by enzymatic degradation;

2) lipid my be removed by two-phase extraction with a hydrophobic solvent;

3) DNA removal is accomplished by enzyme degradation or differential solubility or by centrifugation using dense CsCl solutions;

4) nucleotides and nucleosides are removed by ethanol precipitation or gel filtration chromatography.

10.2

To avoid degradation of RNA by RNase the following precautions should be followed: cleanliness, wearing gloves, autoclaved glassware, diethyl pyrocarbonate treatment, use of detergents, storage of RNA as an ethanol precipitate.

10.3

1) The concentration of nucleic acid is approximately $27 \, \mu g \, cm^{-3}$ (ie 0.6 x 45).

2) d), it is not possible to distinguish DNA and RNA from an absorbance spectrum.

3) We could use the orcinol test for RNA and the diphenylamine reaction for DNA (see Section 7.3 if you do not remember the diphenylamine reaction-Burton method for measuring DNA).

Responses to Chapter 11 SAQs

11.1   1) The methods used for RNA extraction are: detergent/phenol extraction, guanidinium salt technique and the protease K method. In the first two of these methods proteins are removed by denaturation and centrifugation; in the third method proteins are digested. Lipids are removed by solvent extraction. Small molecules can be separated by ethanol precipitation and centrifugation or by gel filtration.

   2) Carbohydrates are removed by centrifugation through sodium acetate and DNA is removed by CsCl density gradient centrifugation or enzyme digestion.

11.2   The absorption maximum is 260 nm. For the general shape of the spectrum refer to Figure 7.3.

11.3   1) The products of the restriction enzyme digestion are linear.

   2) Supercoiled DNA has the highest mobility, linear is next and open circle has the lowest mobility.

11.4   1) An *in vitro* protein synthesis system is a relatively crude cell extract used to synthesise proteins *in vitro*, directed by added mRNA. It is used to characterise mRNA preparations; it is an assay for specific mRNA.

   2) The extract contains tRNA, amino-acyl-tRNA synthetases, ribosomes and all necessary protein factors. Amino acids, an energy supply, (ATP and GTP) and of course mRNA, all need to be supplied.

11.5   1) The sort of figure we hoped you would draw would look like this. The exact spots you choose to shade in each autoradiogram are rather unimportant. What is important is that the spots which appear after a short exposure will become larger on longer exposure.

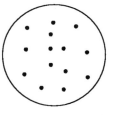   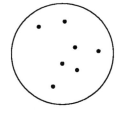   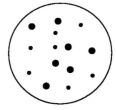

membrane filter   X-ray film with   X-ray film with
                       short exposure   longer exposure

Filter and autoradiograms after development.

   2) Approximately half the mRNA codes for globin.

   3) Approximately half the colonies contain cDNA made from globin mRNA.

4) About half the colonies will shown up quickly, the others much more slowly. this is because the probe is ~50% globin sequence, all other sequences will be represented as a very small fraction and therefore be at a much lower concentration in the hybridisation reaction This is only true if no other mRNA was present in large quantities in the original red blood cells from which the cDNA library was prepared.

11.6

	Fraction method	Advantages	Disadvantages
1)	solubility	simple	poor resolution
2)	sucrose gradient fractionation	good preparation technique	poor resoltuion
3)	agarose gel electrophoresis	excellent resolution	small quantities, therefore limited as a preparative technique
4)	chromatography		
	gel exclusion	good for separating high from low RMM RNA	different mRNA may have similar sizes
	ion exchange	good for separating tRNA	high molecular mass RNA binds too tightly
	HPLC	good resolution	expense of equipment, availability of suitable columns
	affinity	very specific, good for Poly(A)RNA	requires knowledge of RNA sequence

Responses to Chapter 12 SAQs

12.1 1) A denaturing gel electrophoresis system is required because RNA forms secondary structures that have an effect on the mobility during electrophoresis.

2) Formamide or glyoxal are good denaturing agents; the former denatures H-bonds and the latter modifies guanine bases and prevents annealing of complementary regions.

Denatured RNA runs more slowly through the gel than native RNA, (ie has a lower mobility) because denatured RNA is less compact.

12.2 The technique required for determining the RMM of mRNA is Northern blotting. Total RNA is isolated from the cells. It is then separated by denaturing gel electrophoresis and transferred to a nylon membrane. This is probed with a labelled cDNA to the mRNA. A band should appear on autoradiography. Measuring the position of this band in relation to markers of known RMM run on the same gel will reveal the RMM.

12.3 The sequence of RNA in Figure 12.6 is:

GCCGCCUAUGGGGUUAGGCAAUGGAAAUGGCAGCACCC.

12.4 1) a) The base composition of RNA can be determined by alkaline or enzymatic hydrolysis followed by chromatography (eg HPLC or TLC) against known standards. Sensitivity is improved by analysing radioactive mixtures.

b) There are three overall approaches to determining the nucleotide sequence of RNA:

i) endonuclease digestion and two-dimensional separation followed by detailed analysis of individual fragments by repeated end-terminal digestion;

ii) end-labelling and chemical cleavage as in the Maxam and Gilbert technique for DNA sequencing;

iii) primer extension using dideoxynucleoside triphosphates and a radioactive label.

Methods ii) and iii) both involve acrylamide gel electrophoresis to resolve RNA products.

2) If the RNA contains modified nucleotides idiosyncratic results will be observed. This will give a clue to the presence of modified nucleotides. Detailed chemical analysis of digestion products will reveal the nature of the modifications.

Suggestions for further reading

Quick reference dictionaries

J Coombs, Macmillan Dictionary of Biotechnology, Macmillan, 1983

D B Hibbert and A M James, Macmillan Dictionary of Chemistry, Macmillan, 1987

E A Martin, Macmillan Dictionary of Life Sciences, Macmillan, 1985

Standard Texts

D Freifelder, Physical Biochemistry, Applications to Biochemistry and Molecular Biology, Freeman, 1976

K Wilson and K L Goulding, A Biologist's Guide to Principles and Techniques of Practical biochemistry, 3rd Edition, Arnold, 1986

T S Work and E Work, Laboratory Techniques in Biochemistry and Molecular Biology, North-Holland, 7, 1979

Special Texts

Chapter 1

E Sim, Outline Studies in Biology: Membrane Biochemistry, Chapman & Hall, 1982

D Rickwood, Centrifugation: A Practical Approach, Information Retrieval Ltd, 1978

S Lindsay, High Performance Liquid Chromatography, ACOL Series, Wiley, 1987

M Melvin, Electrophoresis, ACOL Series, Wiley, 1987

P A Sewell and B Clarke, Chromatographic Separations, ACOL Series, Wiley, 1987

R C Denney and R Sinclair, Visible and Ultraviolet Spectroscopy, ACOL Series, Wiley, 1987

Chapters 2 to 5

J M Walker (Ed), Methods in Molecular Biology III, New Protein Techniques, Humana Press, 1988

A Darbre, Practical protein Chemistry, A Handbook, Wiley, 1986

E L V Harris and S Angal (Eds), Protein Purification Methods, a Practical Approach, IRL Press, 1989

R Scopes, Protein Purification, Principles and Practice, Springer-Verlag, 1981

Chapters 6 to 11

J M Walker (Ed), Methods in Molecular Biology, IV, New Nucleic Acid Techniques, Humana Press, 1988

R L P Adams, J T Knowler and D P Leader, The Biochemistry of The Nucleic Acids, 10th Ed, Chapman and Hall, 1986

The Molecular Fabric of Cells, BIOTOL Series, Butterworth-Heinemann, 1991

Old R W and Primrose S B, Principles of Gene Manipulation, 3rd Edition, Blackwell Publishers, Oxford, 1985

Appendix 1

Abbreviations used for the common amino acids

Amino acid	Three-letter abbreviation	One-letter symbol
Alanine	Ala	A
Arginine	Arg	R
Asparagine	Asn	N
Aspartic acid	Asp	D
Asparagine or aspartic acid	Asx	B
Cysteine	Cys	C
Glutamine	Gln	Q
Glutamic acid	Glu	E
Glutamine or glutamic acid	Glx	Z
Glycine	Gly	G
Histidine	His	H
Isoleucine	Ile	I
Leucine	Leu	L
Lsyine	Lys	K
Methionine	Met	M
Phenylalanine	Phe	F
Proline	Pro	P
Serine	Ser	S
Threonine	Thr	T
Tryptophan	Trp	W
Tyrosine	Tyr	Y
Valine	Val	V

Appendix 2 Abbreviations and nomenclature of nucleic acids and derivatives

The abbreviations employed in this book are based on those proposed by the Commission on biochmeical Nomenclature (CBN) of the International Union of Pure and Applied Chemistry (IUPAC) and the International Union of Biochemistry (IUB).

Nucleosides

A	adenosine
G	guanosine
C	cytidine
U	uridine
ψ	5-ribosyluracil (pseudouridine)
I	inosine
X	xanthine
rT	ribosylthymine (ribothymidine)
N	unspecified nucleoside
R	unspecified purine nucleoside
Y	unspecified pyrimidine nucleoside
dA	2'-deoxyribosyladenine
dG	2'-deoxyribosylguanine
dC	2'-deoxyribosylcytosine
dT or T	2'-deoxyribosylthymine (thymidine)

Nucleotides

AMP	adenosine 5'-monophosphate
GMP	guanosine 5'-monophosphate
CMP	cytidine 5'-monophosphate
UMP	uridine 5'-monophosphate
dAMP	2'-deoxyribosyladenine 5'-monophosphate
dGMP	2'-deoxyribosylguanine 5'-monophosphate
dCMP	2'-deoxyribosylcytosine 5'-monophosphate
dTMP	2'-deoxyribosylthymine 5'-monophosphate

2'-AMP, 3'-AMP, 5'-AMP etc	2'-, 3'-, 5'-phosphates of adenosine etc
ADP etc	5'-(pyro)diphosphates of adenosine etc
ATP etc	5'-(pyro)triphosphates of adenosine etc
ddTTP etc	2', 3'-dideoxyribosylthymine 5'-triphosphate
araCTP	1-β-D-arabinofuranosylcytosine 5'-triphosphate

Polynucleotides

DNA	deoxyribonucleic acid
cDNA	complimentary DNA (or copy DNA)
mtDNA	mitochondria DNA
RNA	ribonucleic acid
mRNA	messenger RNA
rRNA	ribosomal RNA
tRNA	transfer RNA
nRNA	nuclear RNA
hnRNA	heterogenous nuclear RNA
snRNA	small nuclear RNA
Alanine tRNA or tRNA^{Ala}	transfer RNA that normally accepts alanine
Ala-tRNA^{Ala} or Ala-tRNA	transfer RNA that normally accepts alanine with alanine residue covalently linked

Miscellaneous

RNase, DNase	ribonculease, deoxyribonuclease
P_i, PP_i	inorganic orthophosphate and pyrophosphate
nt	nucleotide
bp	base pair
mt	mitochondrial
cp	chloroplast

Appendix 3

Units of measurement

For historical reasons a number of different units of measurement have evolved. The literature reflects these different systems. In the 1960s many international scientific bodies recommended the standardisation of names and symbols and a universally accepted set of units. These units, SI units (Systeme Internationale de Unites) were based on the definition of: metre (m), kilogram (kg); second (s); ampare (A); mole (mol) and candela (cd). Although, in the intervening period, these units have been widely adopted, their adoption has not been universal. This is especially true in the biological sciences.

It is, therefore, necessary to know both the SI units and the older systems and to be able to interconvert between both sets.

The BIOTOL series of texts predominantly uses SI units. However, in areas of activity where their use is not common, other units have been used. Tables 1 and 2 below provides some alternative methods of expressing various physical quantities. Table 3 provides prefixes which are commonly used.

Mass (SI unit: kg)	Length (SI unit: m)	Volume (SI unit: m^3)	Energy (SI unit: $J = kg\ m^2\ s^{-2}$)
$g = 10^{-3}\,kg$	$cm = 10^{-2}\,m$	$l = dm^3 = 10^{-3}\,m^3$	$cal = 4.184\ J$
$mg = 10^{-3}\,g = 10^{-6}\,kg$	$Å = 10^{-10}\,m$	$dl = 100\ ml = 100\ cm^3$	$erg = 10^{-7}\ J$
$\mu g = 10^{-6}\,g = 10^{-9}\,kg$	$nm = 10^{-9}\,m = 10Å$	$ml = cm^3 = 10^{-6}\,m^3$	$eV = 1.602 \times 10^{-19}\ J$
	$pm = 10^{-12}\,m = 10^{-2}\,Å$	$\mu l = 10^{-3}\,cm^3$	

Table 1 Units for physical quantities

Concentration (SI units: mol m^{-3})

a) $M = mol \, l^{-1} = mol \, dm^{-3} = 10^3 \, mol \, m^{-3}$

b) $mg1^{-1} = \mu g \, cm^{-3} = ppm = 10^{-3} \, g \, dm^{-3}$

c) $\mu g \, g^{-1} = ppm = 10^{-6} \, g \, g^{-1}$

d) $ng \, cm^{-3} = 10^{-6} \, g \, dm^{-3}$

e) $ng \, dm^{-3} = pg \, cm^{-3}$

f) $pg \, g^{-1} = ppb = 10^{-12} \, g \, g^{-1}$

g) $mg\% = 10^{-2} \, g \, dm^{-3}$

h) $\mu g\% = 10^{-5} \, g \, dm^{-3}$

Table 2 Units for concentration

Fraction	Prefix	Symbol	Multiple	Prefix	Symbol
10^{-1}	deci	d	10	deka	da
10^{-2}	centi	c	10^2	hecto	h
10^{-3}	milli	m	10^3	kilo	k
10^{-6}	micro	μ	10^6	mega	M
10^{-9}	nano	n	10^9	giga	G
10^{-12}	pico	p	10^{12}	tera	T
10^{-15}	femto	f	10^{15}	peta	P
10^{-18}	atto	a	10^{18}	exa	E

Table 3 Prefixes for S1 units

Appendix 4

Chemical Nomenclature

Chemical nomenclature is quite a difficult issue especially in dealing with the complex chemicals of biological systems. To rigidly adhere to a strict systematic naming of compounds such as that of the International Union of Pure and Applied Chemistry (IUPAC) would lead to a cumbersome and overly complex text. BIOTOL has adopted a pragmatic approach by predominantly using the names or acronyms of chemicals most widely used in biologically-based activities. It is recognised however that there remains some potential for confusion amongst readers of different background. For example the simple structure CH_3COOH can be described as ethanoic acid or acetic acid depending on the environment or industry in which the compound is produced or used. To reduce such confusion, the BIOTOL series makes every effort to provide synonyms for compounds when they are first mentioned and to provide chemical structures where clarity and context demand.

Index

A

α-amino acid
  general formula, 3
absorbance
  260 nm, 9
  calculation of, 24
  nucleic acid, 6
absorbance at 220 nm, 5
absorbance at 260 nm
  hyperchromic effect, 7
absorbance at 280 nm
  protein estimation, 3
absorbance at 280 nm , 2
absorbance ratio
  A260/A280, 8
acrylamide, 11
  See also polyacrylamide
  polymerisation of, 11
adsorption chromatography, 9
affinity
  See also affinity chromatography
affinity chromatography, 16
  ligand, 20
  ligate, 20
  mRNA purification, 8
  principles of, 17 , 21
  protein purification, 20
  use of spacer arm, 20
agarose, 4
  gel electrophoresis, 6
agarose gel
  electrophoresis apparatus, 5
agarose gel electrophoresis
  RNA, 6
alkaline phosphatase, 10
amino acid
  ionisation of, 6
  R group, 3 , 4
  titration curve, 6
amino acid analysis, 5 , 7 , 9 , 10
  asparagine/aspartic acid, 8
  glutamine/glutamic acid, 8
  ion exchange chromatography, 5
  method of, 10
  tryptophan, 8
amino acid staining
  ninhydrin, 9
amino acids
  reaction with o-phthalaldehyde, 10

B

ammonium sulphate, 23
  precipitation of proteins, 11
  salting out, 14
ampholytes
  carrier, 22
ampicillin, 14
anion exchange chromatography, 13
anion exchanger, 12 , 18
  DEAE cellulose, 18
asparagine/aspartic acid, 8
autoradiography
  use following hybridisation, 12
Avogadro constant, 24

B

base pairing, 2 , 6
Beer-Lambert law, 4 , 25
  protein determination, 4
biological activity, 3
Biuret
  peptide bond, 7
  protein determination, 7
Biuret reagent, 7
Bradford
  colorimetric protein determination, 8
Burton assay, 9

C

caesium chloride, 5
calorimetric estimation of protein, 7
carrier ampholytes, 22
casium chloride
  See CsCl
cation exchanger, 12 , 18
  CM cellulose, 18
cDNA (copy DNA), 12
cDNA library, 10 , 12
cDNA synthesis, 8 , 11
centrifugation, 3
  differential, 3
  isopycnic, 4
  rate-zonal, 4
  ultra, 6
chelating agent, 3
chemical cleavage
  cyanogen bromide, 16
chloroplast DNA, 4
chromatographic bed
  definition of, 8
chromatographic techniques, 8
  See chromatography